T0135583

Mathematical Modelling of
the β-TrCP-dependent Regulation of
Canonical NF-κB and Wnt/β-catenin Signalling.

DISSERTATION

zur Erlangung des akademischen Grades

doctor rerum naturalium

(Dr. rer. nat.)

im Fach Biophysik

eingereicht an der

Mathematisch-Naturwissenschaftlichen Fakultät I

Humboldt-Universität zu Berlin

von

Dipl.-Biochem. Uwe Benary

Präsident der Humboldt-Universität zu Berlin: Prof. Dr. Jan-Hendrik Olbertz

Dekan der Mathematisch-Naturwissenschaftlichen Fakultät I: Prof. Stefan Hecht, Ph.D.

Gutachter/in: 1. Prof. Dr. Dr. h.c. Edda Klipp

             2. Prof. Kwang-Hyun Cho, Ph.D.

             3. Prof. Dr. Olaf Wolkenhauer

Tag der mündlichen Prüfung: 23. Januar 2014

Bibliographic information published by the Deutsche Nationalbibliothek

The Deutsche Nationalbibliothek lists this publication in the Deutsche
Nationalbibliografie; detailed bibliographic data are available
in the Internet at http://dnb.d-nb.de .

©Copyright Logos Verlag Berlin GmbH 2014
All rights reserved.

ISBN 978-3-8325-3732-6

Logos Verlag Berlin GmbH
Comeniushof, Gubener Str. 47,
10243 Berlin
Tel.: +49 (0)30 42 85 10 90
Fax: +49 (0)30 42 85 10 92
INTERNET: http://www.logos-verlag.de

Zusammenfassung

Eine Zelle erhält Informationen über ihre Umgebung und ihren intrazellulären Zustand durch Signale, die von spezifischen Signalrezeptoren über Signalwege in den Zellkern übermittelt werden. Dort regulieren die verschiedenen aufeinandertreffenden Signale die zelluläre Genexpression, so dass die Zelle angemessen auf Veränderungen ihrer Umgebung und ihres intrazellulären Zustands reagieren kann. Eine gestörte Signalübertragung kann zu fehlgeleiteten zellulären Reaktionen führen, zum Beispiel zu einer unkontrollierten Vermehrung der Zelle wie bei Krebserkrankungen. Signalwege bestehen aus komplexen Wechselwirkungen zwischen vielen Signalmolekülen. Diese Wechselwirkungen können regulatorische Rückkopplungsmechanismen in den Signalwegen erzeugen oder zu Informationsaustausch zwischen Signalwegen über gemeinsam genutzte Signalwegskomponenten führen. Auf diese Weise bilden Signalwege komplexe Netzwerke, die ein genaueres Verständnis darüber, wie ein Signal durch das Netzwerk der Signalwege übermittelt wird, erschweren. Darüber hinaus ist es schwierig, geeignete Stellen in diesen Netzwerken vorherzusagen, an denen korrigierend in eine gestörte Signalübermittlung eingegriffen werden könnte. Die mathematische Modellierung der molekularen Prozesse bei der Signalübertragung stellt eine Methode dar, nützliche Erkenntnisse bezüglich der Regulation der Signalübermittlung zu gewinnen. Die vorliegende Dissertation konzentriert sich auf die Entwickelung und Analyse mathematischer Modelle des kanonischen NF-κB-Signalweges und des Wnt/β-catenin-Signalweges.

Der kanonische NF-κB-Signalweg und der Wnt/β-catenin-Signalweg regulieren viele zentrale Prozesse in der Zelle, wie zum Beispiel Zellproliferation, Zelldifferenzierung und Zelltod. Der kanonische NF-κB-Signalweg und der Wnt/β-catenin-Signalweg sind bekannte Beispiele für Signalwege, deren Aktivität durch die Ubiquitinierung mit anschließendem proteasomalen Abbau zentraler Signalwegskomponenten gesteuert wird. Im Fall des Wnt/β-catenin bzw. kanonischen NF-κB-Signalwegs werden β-catenin bzw. der NF-κB-Inhibitor IκB kovalent mit dem kleinen Protein Ubiquitin markiert, so dass IκB und β-catenin gezielt im Proteasom abgebaut werden können. Die Ubiquitinierung von IκB sowie β-catenin wird durch zwei Paraloge des *β-transducin repeat-containing* Proteins (β-TrCP) vermittelt. Die Expression der beiden β-TrCP-Paraloge wird durch den Wnt/β-catenin-Signalweg beeinflusst, wodurch zwei unterschiedliche Rückkopplungsmechanismen im Wnt/β-catenin-Signalweg entstehen. Der Wnt/β-catenin-Signalweg könnte außerdem durch die transkriptionale Regulation der β-TrCP-Paraloge auf die Signalübertragung des kanonischen NF-κB-Signalwegs einwirken. Zusätzlich könnten sich der kanonischen NF-κB-Signalweg und der Wnt/β-catenin-Signalweg auch durch ihre Konkurrenz um β-TrCP gegenseitig beeinflussen, da die Aktivität beider Signalwege von β-TrCP abhängt.

In der vorliegenden Dissertation wird der aktuelle biologische Erkenntnisstand zu β-TrCP zusammengefasst und in der Weiterentwicklung mathematischer Modelle des kanonischen NF-κB-Signalwegs und des Wnt/β-catenin-Signalwegs berücksichtigt. Die entwickelten mathematischen Modelle verwenden gewöhnliche Differentialgleichungen, um die zeitlichen Veränderungen in den Konzentrationen der Signalwegskomponenten beider Signalwege zu beschreiben. Anhand dieser mathematischen Modelle werden die möglichen Folgen veränderter β-TrCP-Konzentrationen auf die Signalübertragung im kanonischen NF-κB und Wnt/β-catenin-Signalweg untersucht. Im Falle des Wnt/β-catenin-Signalwegs wird der Einfluss der β-TrCP-vermittelten Rückkopplungsmechanismen auf die Signalübermittlung im Signalweg analysiert. Darüber hinaus werden die Bedingungen untersucht, die eine mögliche Wechselwirkung zwischen kanonischem NF-κB und Wnt/β-catenin-Signalweg aufgrund ihrer beider Konkurrenz um β-TrCP ermöglichen oder verhindern.

Die Analyse des entwickelten Wnt/β-catenin-Signalweg-Modells zeigt, dass die beiden β-TrCP-Paraloge unterschiedliche Funktionen im Wnt/β-catenin-Signalweg erfüllen. Die Simulationen des Modells zeigen, dass ein Paralog einen starken Einfluss auf die Signalübertragung im Wnt/β-catenin-Signalweg hat, während das zweite Paralog einen Absicherungsmechanismus gegen mutationsbedingte Funktionseinbußen des anderen Paralogs darstellt. Die Ergebnisse stellen somit die bisherige Annahme einer funktionellen Redundanz der β-TrCP-Paraloge im Wnt/β-catenin-Signalweg in Frage. Die Analyse legt nahe, dass die tatsächlichen Auswirkungen auf die Signalübermittlung im Wnt/β-catenin-

Signalweg vor allem von den jeweils vorhandenen Konzentrationen der Paraloge abhängt. Die Untersuchung des mathematischen Modells des kanonischen NF-κB-Signalwegs zeigt, dass eine transkriptionale Regulation der β-TrCP-Konzentration auch die Signalübertragung im kanonischen NF-κB-Signalweg beeinflusst. Dieses Ergebnis weist auf eine mögliche Beeinflussung der Signalübertragung des kanonischen NF-κB-Signalwegs durch den Wnt/β-catenin-Signalweg mittels transkriptionaler Regulation der β-TrCP-Paraloge hin. Unter Verwendung eines Minimalmodells, das beide Signalwege vereint, aber diesen transkriptionalen Regulationsmechanismus nicht berücksichtigt, werden die Bedingungen untersucht, unter denen Wechselwirkungen zwischen kanonischem NF-κB und Wnt/β-catenin-Signalweg aufgrund ihrer Konkurrenz um β-TrCP ermöglicht oder verhindert werden. In den Simulationen des Minimalmodells wird nur eine Einflussnahme des kanonischen NF-κB-Signalwegs auf die Signalübermittlung des Wnt/β-catenin-Signalwegs beobachtet. Die Analyse zeigt, dass diese Wechselwirkung durch die Parameter am stärksten beeinflusst wird, die den β-TrCP-vermittelten Abbau von IκB und die Produktion bzw. den Abbau von β-TrCP beschreiben. Eine geringe β-TrCP-Konzentration scheint hingegen keine ausreichende Bedingung zu sein, um Wechselwirkungen mittels konkurrierender Bindung von β-TrCP zwischen den beiden Signalwegen zu ermöglichen.

Die gewonnenen Erkenntnisse aus den Modellanalysen heben vor allem die Bedeutung der Konzentration der einzelnen β-TrCP-Paraloge in der Regulation des kanonischen NF-κB und Wnt/β-catenin-Signalwegs hervor. Sie schlagen alternative mechanistische Erklärungen für widersprüchliche experimentelle Beobachtungen vor, die sich auf die Auswirkungen des Wnt/β-catenin-Signalwegs auf die Signalübertragung des kanonischen NF-κB-Signalwegs beziehen. In den Experimenten scheint der Wnt/β-catenin-Signalweg je nach untersuchtem Zelltyp entweder die Signalübermittlung des kanonischen NF-κB-Signalwegs zu verstärken oder zu hemmen. Die Modellsimulationen zeigen, dass die Konzentration der einzelnen β-TrCP-Paraloge in diesem Fall eine zentrale Rolle spielen könnte.

Abstract

A cell gains information about its environment and intracellular state through signals that are transduced from specific signal receptors via signalling pathways into the nucleus of the cell. In the nucleus, the different signals are integrated to regulate gene expression enabling the cell to adequately react to environmental and/or intracellular changes. Aberrant signal transduction can result in inappropriate cellular responses, for instance, uncontrolled proliferation as in cancer. Signalling pathways are built of complex interactions between many signalling molecules. These interactions may create regulatory feedback mechanisms in a signalling pathway and/or may allow for communication between signalling pathways via shared pathway components. In this way, signalling pathways form complex signalling networks that render it very difficult to understand how a signal propagates through the network. Moreover, it is difficult to predict interference strategies to correct for aberrant signal transduction. Mathematical modelling of the molecular processes of signal transduction provides a method to gain deeper insight into the regulation of signal transduction. This thesis focusses on the development and analysis of mathematical models of canonical NF-κB and Wnt/β-catenin signal transduction.

The canonical NF-κB signalling pathway and the Wnt/β-catenin signalling pathway regulate many central processes in the cell such as cell proliferation, differentiation, and survival. They are well-known examples of signal transduction pathways whose activity is controlled by kinase-induced ubiquitination and subsequent proteasomal degradation of central pathway components. In the case of canonical NF-κB and Wnt/β-catenin signalling, the inhibitor of NF-κB (IκB) and β-catenin, respectively, are covalently modified with the small protein ubiquitin, which labels IκB and β-catenin for degradation in a timely and selective manner. Ubiquitination of IκB as well as β-catenin is mediated by the β-transducin repeat-containing protein (β-TrCP). β-TrCP is expressed in two paralogues that are both regulated by Wnt/β-catenin signalling. In that way, two different transcriptional feedback mechanisms are established in the Wnt/β-catenin pathway. In addition, the Wnt/β-catenin pathway may influence canonical NF-κB signalling via transcriptional regulation of β-TrCP. Furthermore, the canonical NF-κB and Wnt/β-catenin signalling pathway may interfere (crosstalk) with each other via competitive β-TrCP sequestration since both pathways equally depend on the availability of β-TrCP.

In this thesis, the current biological knowledge regarding β-TrCP is collected and integrated into a mathematical framework of canonical NF-κB and Wnt/β-catenin signalling. The applied mathematical models use ordinary differential equations to account for the temporal changes in the concentrations of components of the canonical NF-κB and Wnt/β-catenin signalling pathway. Using these mathematical models, the possible consequences of changes in β-TrCP abundance on Wnt/β-catenin and canonical NF-κB signal transduction are investigated. In the case of Wnt/β-catenin signalling pathway, the impact of the β-TrCP-mediated feedback mechanisms on Wnt/β-catenin signal transduction is analysed. Furthermore, the conditions are explored that enable and prevent potential crosstalk via competitive β-TrCP sequestration between canonical NF-κB and Wnt/β-catenin signalling.

The analysis of the extended model of Wnt/β-catenin signalling demonstrates that the two paralogues of β-TrCP fulfil distinct functions in Wnt/β-catenin signalling. The simulations of the model show that one paralogue influences Wnt/β-catenin signalling dynamics, while the second paralogue may serve as a back-up mechanism for the case of loss-of-function mutations in the first paralogue. These results question the current notion of functional redundancy of the β-TrCP paralogues in Wnt/β-catenin signalling. The analysis shows that the effective impact of the paralogues of β-TrCP on Wnt/β-catenin signalling depends on the actual expression levels of each paralogue. The analysis of the model of canonical NF-κB signalling demonstrates that transcriptional regulation of β-TrCP abundance influences the transient dynamics as well as the stimulated steady state concentration of nuclear NF-κB. This result indicates that Wnt/β-catenin signalling could potentially affect canonical NF-κB activation through regulation of β-TrCP abundance. In a minimal model of combined canonical NF-κB and Wnt/β-catenin signalling, which neglects this transcriptional regulation mechanism, the conditions supporting or preventing crosstalk by competitive β-TrCP sequestration between the pathways are explored. In the simulations, crosstalk by competitive β-TrCP sequestration is only observed from the canonical NF-κB to the Wnt/β-catenin signalling pathway. The analysis reveals that this crosstalk is influenced by the parameters associated with β-TrCP-mediated IκB degradation in combination with those of β-TrCP production and degradation. Low abundance of β-TrCP is however no sufficient condition to enable crosstalk via competitive β-TrCP sequestration.

The insights gained in the analyses of the mathematical models emphasise the critical role of the actual abundance of each β-TrCP paralogue on the regulation of canonical NF-κB and Wnt/β-catenin pathway activity. The insights offer alternative mechanistic explanations to account for conflicting experimental observations concerning the impact of Wnt/β-catenin signalling on canonical NF-κB pathway activity. In the experiments, Wnt/β-catenin signalling seemed to either enhance or inhibit canonical NF-κB signal transduction depending on the particular cell type. The model simulations indicate that the abundance of each β-TrCP paralogue may play a central role in this case.

Keywords: mathematical modelling, β-TrCP, HOS, FWD1, Wnt/β-catenin signalling, NF-κB signalling, feedback, crosstalk, mutation, cancer

Contents

Abbreviations

293T	a human embryonic kidney cell line
Aa	amino acid
APC	adenomatous polyposis coli protein
Asp	aspartic acid
ATF	activating transcription factor
ATP	adenosine-5'-triphosphate
Axin	axis inhibition protein
BMP-2	bone morphogenetic protein 2
CD4	cluster of differentiation 4
CI	crosstalk impact
CK1	casein kinase 1
CP	core particle of the 26S proteasome
CRD-BP	coding region determinant-binding protein
Cul1	cullin 1
DC	destruction complex
DNA	deoxyribonucleic acid
Dsh	Dishevelled
DUBs	Deubiquitinases
E1	ubiquitin-activating enzyme
E2	ubiquitin-conjugating enzyme
E3	ubiquitin ligase
FWD1	also known as β-TrCP1
Gly	Glycine

GSK3	glycogen synthase kinase 3
HB	Hopf bifurcation
HECT	homologous to the E6-AP carboxy-terminus
HIV-1	human immunodeficiency virus 1
HOS	Homologous to Slimb protein
IKK	IκB kinase
IL-1	Interleukin-1
int1	murine integration 1 gene
IκB	inhibitor of NF-κB
JNK	c-Jun N-terminal kinase
LDL	low-density lipoprotein
LEF	lymphoid enhancer-binding factor
LPS	lipopolysaccharide
LRP5/6	low-density lipoprotein (LDL) receptor-related protein 5/6
LZTS2	leucine zipper putative tumour suppressor 2
MAPK	mitogen-activated protein kinases
MHC	major histocompatibility complex
mRNA	messenger RNA
NEMO	NF-κB essential modulator
NF-κB	nuclear factor κ-light-chain-enhancer of activated B cells
NIH3T3	a mouse embryonic fibroblast cell line
ODE	ordinary differential equation
PCP	planar cell polarity pathway
PKB	Protein Kinase B
Rbx1/Roc1	RING-domain protein regulator of cullins 1
RING	really interesting new gene
RIP	receptor interacting protein
RNA	ribonucleic acid
RP	regulatory particle of the 26S proteasome
SCF	Skp-Cullin-F-box complex
Ser	Serine

siRNA	small interfering RNA
Skp1	S-phase kinase-associated protein 1
TCF	T-cell factor/lymphoid enhancer factor family
Thr	Threonine
TNFR1	essential mediator of the proximal TNFα receptor 1
TNFα	tumour necrosis factor α
Ub	Ubiquitin
UV	Ultraviolet
Vpu	viral protein unique
wg	drosophila wingless gene
β-TrCP	β-transducin repeat-containing protein

1. Introduction to the biological scope of the thesis

Signal transduction refers to the intracellular processes that amplify and transmit extracellular stimuli from the plasma membrane into the cell's nucleus to regulate gene expression. The nuclear factor κ-light-chain-enhancer of activated B cells (NF-κB) signalling pathway and the Wnt/β-catenin signalling pathway are two important signal transduction pathways that are involved in the regulation of cell proliferation, differentiation, and survival. In both signalling pathways, ubiquitination of pathway components has evolved as a key regulatory mechanism of pathway activity. Ubiquitination is the covalent modification of a protein by attaching the small protein ubiquitin. Ubiquitination controls activity, abundance, and subcellular localisation of several signalling proteins in the NF-κB and the Wnt/β-catenin signalling pathway. Most notably, the stability of two central signal mediators, inhibitor of NF-κB (IκB) and β-catenin, is controlled by ubiquitination and subsequent degradation in the proteasome. Ubiquitination of IκB as well as β-catenin is mediated by the same β-transducin repeat-containing protein (β-TrCP) offering a potential mechanism through which both signalling pathways may interfere with each other.

This introduction provides an overview of the current scientific knowledge on the function and regulation of β-TrCP, which mediates the ubiquitination of IκB and β-catenin, and thereby links Wnt/β-catenin signalling, NF-κB signalling, and the ubiquitin/proteasome system (Section 1.1 to Section 1.4). At the end of this introduction, three main questions concerning β-TrCP in Wnt/β-catenin and NF-κB signalling are extracted that define the objectives of this thesis (Section 1.5).

1.1 The ubiquitin/proteasome system

Three major protein degradation pathways regulate protein abundance in mammalian cells: the lysosomal pathway, the autophagosomal pathway, and the proteasomal pathway (Clague and Urbe, 2010). Whereas the lysosomal pathway mediates the degradation of membrane-associated proteins, such as receptors or channels, the autophagosomal pathway digests protein aggregates or whole organelles. In contrast to the former two pathways, the proteasomal degradation pathway allows for selective target-specific degradation of proteins. Although the relative contribution of the three pathways may vary between cell types and environmental conditions, it is generally assumed that proteasomal degradation predominates in cells cultured under stress-free conditions (Clague and Urbe, 2010; Deshaies, 1999; DiDonato et al., 2012). Despite differences in target selectivity between the three degradation pathways, they share their dependence on ubiquitination of proteins (Clague and Urbe, 2010).

1.1.1 The process of ubiquitination

Ubiquitin is an ubiquitously expressed 76-amino-acid polypeptide, which is highly conserved throughout eukaryotes (Goldstein et al., 1975; Sharp and Li, 1987). It functions as a post-translational modifier to regulate protein activity and abundance. Ubiquitin is covalently ligated to proteins in a process called ubiquitination.

Ubiquitination occurs through three sequential steps catalysed by ubiquitin-activating, ubiquitin-conjugating, and ubiquitin-ligase enzymes (Deshaies, 1999; Hershko et al., 1983). Ubiquitin is first attached to an ubiquitin-activating enzyme (E1) in an ATP-dependent reaction to form a high-energy thiolester intermediate (Figure 1.1). This ubiquitin is then transferred from the E1 protein to an ubiquitin-conjugating enzyme (E2). Finally, with the assistance of an ubiquitin ligase (E3) ubiquitin is transferred from the E2 enzyme to a lysine residue of the target protein. A repetitive iteration of these three steps leads to the assembly of a polyubiquitin chain.

Ubiquitin ligases can be classified into two main families in eukaryotes (de Bie and Ciechanover, 2011; Hochrainer and Lipp, 2007). First, there is the considerably large family of RING (really interesting new gene) domain-containing E3s with about 600 ligases identified in humans (de Bie and Ciechanover, 2011; Deshaies and Joazeiro, 2009). The second family comprises monomeric HECT (homologous to the E6-AP carboxy terminus) domain-containing E3s enumerating to approximately 30 human enzymes (de Bie and Ciechanover, 2011). This variety of ubiquitin ligases outnumbers by far the two human E1 and fewer than 40 E2 genes (Deshaies and Joazeiro, 2009). HECT E3s contain a cysteine residue that is transiently bound by ubiquitin during its transfer from the E2 enzyme to the target protein (Huibregtse et al., 1995). In contrast, RING ligases merely serve as scaffolds facilitating the direct transfer of ubiquitin (Petroski and Deshaies, 2005).

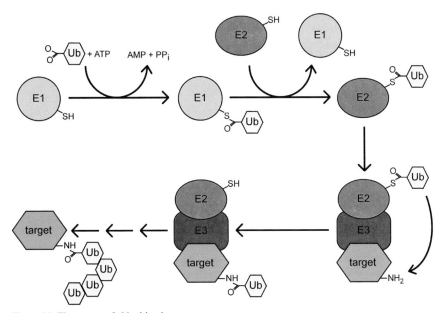

Figure 1.1: The process of ubiquitination.
ATP-dependent activation and transfer of ubiquitin (Ub) onto a target protein by sequential action of an ubiquitin-activating enzyme (E1), an ubiquitin-conjugating enzyme (E2), and an ubiquitin ligase (E3) is shown. Detailed explanations are provided in the text of Section 1.1.1. [1]

Each E3 enzyme binds only one or a few E2 and recognises a restricted set of specific target proteins (Deshaies, 1999; Maniatis, 1999; Ravid and Hochstrasser, 2008). Thus, the involved ubiquitin ligase provides the target selectivity in the ubiquitination process. In consequence, regulation of E3 activity or E3-target interaction has been proposed to serve as the primary mechanism to control protein degradation *in vivo* (de Bie and Ciechanover, 2011; Deshaies, 1999; Ravid and Hochstrasser, 2008).

---

[1]The figure was adapted from the file "Schematic diagram of the ubiquitination system." created by Roger B. Dodd, available under a licensed under the Creative Commons Attribution-Share Alike 3.0 Unported license from http://upload.wikimedia.org/wikipedia/commons/5/5b/Ubiquitylation.svg .

1.1.2 Ubiquitin-dependent proteolysis by the proteasome

The protein-linked polyubiquitin chain of four or more ubiquitin molecules is recognised by the 26S proteasome complex leading to the digestion of the targeted protein. The proteasome is a 2.5 MDa complex consisting of two regulatory particles capping the openings of one core particle (Finley, 2009; Pickart, 2004). The core particle harbours three catalytically active sites in its central cavity (Figure 1.2). Its chymotrypsin-like, trypsin-like, and caspase-like activities cleave the targeted protein into three to 15 amino-acid short oligopeptides (Frankland-Searby and Bhaumik, 2012; Saeki and Tanaka, 2012). These peptides are subsequently hydrolysed to single amino-acids by cytosolic peptidases. In higher eukaryotes, the oligopeptides may also be presented by the major histocompatibility complex (MHC) class I to the immune system (Goldberg, 2007; Saeki and Tanaka, 2012).

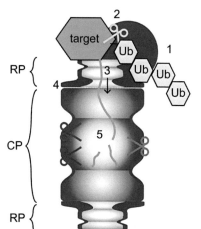

Figure 1.2: Protein degradation by the 26S proteasome. The ubiquitin chain is recognised (1) and cleaved off by associated deubiquitinating enzymes represented by the yellow scissors (2). The target protein is unfolded and translocated into the central cavity of the core particle (3) conducted by ATPase subunits (4). The protein is cleaved by three catalytically active sites in the central cavity represented by three coloured scissors (5). CP: core particle; RP: regulatory particle; Ub: ubiquitin. Figure adapted from (Saeki and Tanaka, 2012).

To excess the proteolytically active sites located in the interior space of the core particle, the polyubiquitinated protein needs to be unfolded and translocated through a narrow channel into the central cavity (Figure 1.2). This well regulated event is conducted in an ATP-dependent manner mediated by the six ATPase subunits of the regulatory particle (Groll et al., 2000; Navon and Goldberg, 2001). Additionally, the regulatory particle is responsible for the recognition of the polyubiquitin chain linked to the protein (Finley, 2009; Frankland-Searby and Bhaumik, 2012). The ubiquitin chain is cleaved off by associated deubiquitinating enzymes prior to the unfolding process allowing for recycling of ubiquitin for reuse in the cell (Finley, 2009; Hershko et al., 1980; Saeki and Tanaka, 2012).

1.1.3   Functions of ubiquitin in addition to protein degradation

Besides its function in all three major protein degradation pathways, ubiquitin ligation fulfils numerous other cellular functions. They range from signal receptor maturation and trafficking to regulation of protein conformation and protein-protein-interactions (Bianchi and Meier, 2009; Clague and Urbe, 2010; de Bie and Ciechanover, 2011; Hochrainer and Lipp, 2007; Iwai, 2012; Kanarek and Ben-Neriah, 2012). Each ubiquitin molecule contains seven accessible lysine residues for further ubiquitin ligation (Hochrainer and Lipp, 2007). In this way, several types of polyubiquitin chains can be generated that differ in chain length and the type of linkage depending on the lysine residue used. It has been hypothesised that the actual type of polyubiquitin chain determines the mode of protein regulation (Iwai, 2012; Schmukle and Walczak, 2012). The modification with ubiquitin is also a reversible process (de Bie and Ciechanover, 2011; Hochrainer and Lipp, 2007). Ubiquitin can be cleaved off from target proteins by deubiquitinases (DUBs). In this way, DUBs oppose the function of E3 ligases.

Ubiquitin has emerged as a central regulatory protein in Wnt/β-catenin and NF-κB signal transduction (Iwai, 2012; Maniatis, 1999; Schmukle and Walczak, 2012; Tauriello and Maurice, 2010). In both pathways, protein modifications by ubiquitin control stability, activity, and function of several crucial signalling components (Chen, 2012; Harhaj and Dixit, 2012; Iwai, 2012; Tauriello and Maurice, 2010). The Wnt/β-catenin and the NF-κB signalling pathways are introduced in more detail in the following sections (Section 1.2 and Section 1.3, respectively).

1.2   The Wnt/β-catenin signalling pathway

The term Wnt is a portmanteau of the abbreviated names of two genes, drosophila wingless (wg) and murine integration 1 (int1), whose discoveries initiated the research of Wnt signal transduction (Nusse and Varmus, 2012). Wnts are a large family of secreted glycoproteins that play crucial roles in embryonic development as well as stem cell homeostasis and cell fate specification in adult tissues (Klaus and Birchmeier, 2008; Merrill, 2012; Willert and Nusse, 2012). Wnt receptor activation is transduced by three different signalling pathways: the planar cell polarity (PCP) pathway, the Wnt/Ca^{2+} pathway, and the Wnt/β-catenin pathway (Kuhl et al., 2000; McNeill and Woodgett, 2010; van Amerongen et al., 2008; Veeman et al., 2003). The Wnt/β-catenin pathway is currently the best characterised of these three. A central task in the Wnt/β-catenin signalling pathway is the control of β-catenin concentration and activity (Stamos and Weis, 2013). β-Catenin, which was originally described as a component of adherent junctions in complex with E-cadherin (Archbold et al., 2012; Valenta et al., 2012), functions in the context of Wnt signalling as a transcriptional regulator of Wnt/β-catenin target genes (Harris and Peifer, 2005; MacDonald et al., 2009).

In resting cells, the concentration of unbound β-catenin remains low because its permanent production is counterbalanced by its continuous degradation. The degradation is primarily mediated by a destruction complex (Figure 1.3A). In this complex, axis inhibition protein (Axin) and adenomatous polyposis coli protein (APC) form a scaffold that allows casein kinase 1 (CK1) and glycogen synthase kinase 3 (GSK3) to sequentially phosphorylate β-catenin at specific serine and threonine residues (Ser45, Thr41, Ser37, and Ser33) (Doble and Woodgett, 2003; Hart et al., 1998). Phosphorylated β-catenin is recognised by β-TrCP, ubiquitinated and subsequently degraded by the 26S proteasome (Aberle et al., 1997; Fuchs et al., 1999; Hart et al., 1999; Liu et al., 1999).

Wnt ligands stimulate cells by binding to the cell surface receptor Frizzled and its co-receptors low-density lipoprotein (LDL)-related protein 5/6 (LRP5/6) (MacDonald and He, 2012; Niehrs, 2012) (Figure 1.3B). Subsequently, the signal mediator Dishevelled (Dsh) is activated and promotes, by a yet unresolved mechanism, the dissociation of the destruction complex (Kimelman and Xu, 2006; Stamos and Weis, 2013). As a result, β-catenin accumulates in the cytosol and translocates into the nucleus, where it interacts with transcription factors of the T-cell factor/lymphoid enhancer factor (TCF) family (Cadigan and Waterman, 2012). β-Catenin/TCF complexes regulate the expression of Wnt/β-catenin target genes, such as cyclin D1, c-myc, and Axin2 (Giles et al., 2003; He et al., 1998; Shtutman et al., 1999; Stadeli et al., 2006). A frequently updated list of further Wnt/β-catenin target genes can be found in the internet at http://www.stanford.edu/group/nusselab/cgi-bin/wnt. In addition to β-catenin/TCF, co-regulators (Figure 1.3) and their posttranslational modifications have been suggested to contribute to the regulation of Wnt/β-catenin target gene expression (Archbold et al., 2012; Hecht and Kemler, 2000).

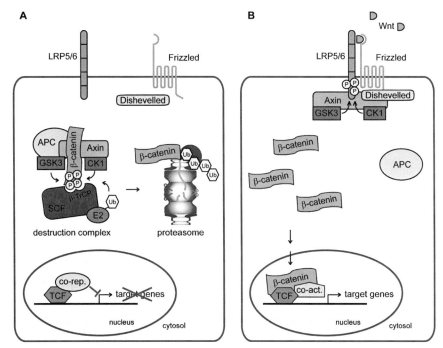

Figure 1.3: The Wnt/β-catenin signalling pathway.

(A) In unstimulated cells, constitutive production of β-catenin is counterbalanced by its continuous degradation by the proteasome. The degradation of β-catenin is initiated by its phosphorylation and subsequent recognition by β-TrCP that mediates β-catenin ubiquitination. In that way, β-catenin concentrations are kept at low levels. (B) Wnt stimulation results in Dishevelled-mediated recruitment of Axin, GSK3, and CK1 to the receptors Frizzled and LRP5/6. Consequently, the destruction complex dissociates leading to impaired β-catenin degradation and thus β-catenin concentration increases. β-Catenin binds TCF to regulate target gene expression. Co-act: co-activators; co-rep: co-repressors; E2: ubiquitin-conjugating enzyme; P: phosphate; SCF: Skp-Cullin-F-box complex (see Section 1.4.1); Ub: ubiquitin.

Aberrant activation of Wnt/β-catenin signalling is associated with many human diseases such as neurodegenerative diseases (Clevers and Nusse, 2012; Inestrosa and Arenas, 2010; MacDonald et al., 2009). Mutations in pathway components were frequently found in various types of cancer (Clevers, 2006; Giles et al., 2003). For instance, in more than 85% of colon cancer samples mutations in APC have been detected (Giles et al., 2003).

1.3   The canonical NF-κB signalling pathway

Nuclear factor κ-light-chain-enhancer of activated B cells (NF-κB) signal transduction plays a fundamental role in the regulation of inflammation and immunity and is also involved in many key cellular processes such as cell proliferation, differentiation, and survival (Ben-Neriah and Karin; Hayden and Ghosh, 2012). Thus, it is not surprising that aberrant regulation of NF-κB activation may lead to severe diseases such as arthritis or autoimmunity. Up-regulated NF-κB activity has been detected in various human cancers where it can promote cell resistance to anti-cancer therapy (Bhatia et al., 2002; DiDonato et al., 2012).

NF-κB is a collective designation for a family of inducible, sequence-specific transcription factors. The mammalian NF-κB transcription factor family consists of five proteins: p65 (RelA), RelB, c-Rel, precursor protein p105 and its cleavage product p50 (NF-κB1), as well as the precursor protein p100 and its cleavage product p52 (NF-κB2). All family members associate with each other to form homo- and heterodimeric complexes that fulfil presumably distinct functions (O'Dea and Hoffmann, 2010; Sen and Smale, 2010). They all share a conserved 300 amino acid long N-terminal Rel homology domain which is required for dimerisation, interaction with the inhibitors of NF-κB (IκBs), nuclear translocation as well as DNA binding. The p50/p65 heterodimer is the most abundant NF-κB dimer found in almost all mammalian cell types (Amit and Ben-Neriah, 2003; Hochrainer and Lipp, 2007; Oeckinghaus and Ghosh, 2009).

In resting cells, NF-κB dimers are retained inactive in the cytosol through their interaction with IκB proteins. IκB represents a family of NF-κB inhibitors: IκBα, IκBβ, IκBε, and the ankyrin-containing Rel proteins (Amit and Ben-Neriah, 2003; Hinz et al., 2012). The p65/p50 heterodimer is primarily associated with IκBα, which is the best-studied member of the IκB family (O'Dea and Hoffmann, 2010).

NF-κB can be activated by many different extracellular stimuli including inflammatory cytokines, such as tumour necrosis factor α (TNFα) and interleukin 1 (IL-1), bacterial lipopolysaccharide (LPS) and viruses, or cellular stresses such as UV-light (Chen, 2012; Pahl, 1999). With few exceptions (for instance, UV radiation) the diverse signalling routes leading to NF-κB activation all converge on subunits of the IκB kinase (IKK) complex, namely IKKα, IKKβ, and NF-κB essential modulator (NEMO) (Amit and Ben-Neriah, 2003; Hayden and Ghosh, 2012; Scheidereit, 2006). Depending on the kind of inducing stimulus, the signal transducing IKK, and the NF-κB dimers that become activated, two NF-κB signalling branches are distinguished: the canonical and non-canonical NF-κB pathway (Hochrainer and Lipp, 2007; Scheidereit, 2006; Sun, 2012). The two pathways also differ in their characteristic response times; namely fast activation of the canonical pathway as

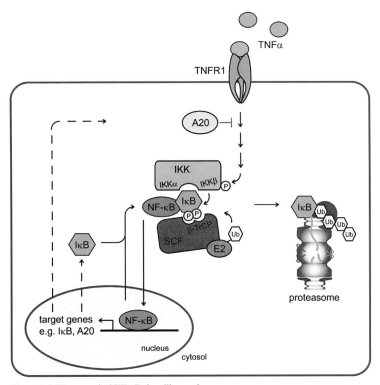

Figure 1.4: The canonical NF-κB signalling pathway.

In resting cells, NF-κB is retained inactive in the cytosol by binding to its inhibitor IκB. Upon receptor activation by TNFα, the pathway activating signal is transduced to activate IKKβ by phosphorylation. IKKβ phosphorylates IκB, which is then recognised by β-TrCP, ubiquitinated, and degraded by the proteasome. IκB degradation releases NF-κB, which translocates into the nucleus to regulate target gene expression. For instance, NF-κB induces the expression of its own inhibitor IκB and the pathway inhibitor A20. IκB inactivates NF-κB transcriptional activity by means of its re-localisation to the cytosol. E2: ubiquitin-conjugating enzyme; IKK: IκB kinase complex; P: phosphate; SCF: Skp-Cullin-F-box complex (see Section 1.4.1); TNFR1: TNFα receptor 1; Ub: ubiquitin.

opposed to slow kinetics of the non-canonical branch (Scheidereit, 2006). This thesis focusses on the canonical NF-κB signalling branch and its activation by TNFα. For the sake of simplicity, the term NF-κB will hereafter refer to the p65/p50 dimer, IκB will refer to IκBα, and TNF is used instead of TNFα, unless stated otherwise.

An important event in the canonical pathway is the activation of IKKβ (Figure 1.4), which leads to the phosphorylation of two N-terminal serine residues (Ser32 and Ser36) of IκB (Alkalay et al., 1995; Chen et al., 1995). Phosphorylated IκB is recognised by β-TrCP, ubiquitinated, and rapidly degraded by the 26S proteasome complex (Fuchs et al., 2004; Hinz et al., 2012). The degradation of IκB allows NF-κB dimers to translocate into the nucleus where they regulate the expression of NF-κB target genes (Henkel et al., 1993). Several target genes encode components of the pathway that possibly establish regulatory feedbacks in the NF-κB response. Examples are IκB and A20 (Figure 1.4). A list of target genes can be found in the internet at www.nf-kb.org.

The new IκB, whose expression is induced by nuclear NF-κB, is assumed to play a major role in terminating the NF-κB response (Oeckinghaus and Ghosh, 2009). In this scenario, IκB binds NF-κB in the nucleus, detaches NF-κB from the DNA, and re-locates it to the cytosol (Figure 1.4). An additional mechanism that may contribute to the termination of signalling involves the NF-κB target gene product A20. A20 is reported to deubiquinate the receptor interacting protein (RIP), an essential mediator of the proximal TNFα receptor 1 (TNFR1) signalling complex (Heyninck and Beyaert, 2005; Wertz et al., 2004). By this process, signal transduction can be interrupted.

1.4   The β-transducin repeat-containing protein family

1.4.1   β-TrCP is a subunit of the Skp-Cullin-F-box ubiquitin ligase complex

Skp-Cullin-F-box (SCF) complexes are multi-subunit members of the RING family of E3s (Hochrainer and Lipp, 2007). The core components of the founding member of the SCF complex family includes the scaffold protein cullin 1 (Cul1), the RING-domain protein regulator of cullins 1 (Rbx1/Roc1), the adaptor protein S-phase kinase-associated protein 1 (Skp1), and an F-box protein that binds the target protein (Deshaies, 1999; Deshaies and Joazeiro, 2009). A schematic illustration of the SCF complex is given in Figure 1.5. The subunits of the SCF complex are highly conserved throughout eukaryotes and fulfil distinct functions (Deshaies, 1999). Rbx1/Roc1 links the scaffold protein Cul1 to the ubiquitin-conjugating enzyme E2 (Skowyra et al., 1999). Skp1 interacts simultaneously with Cul1 and with the F-box motif of the F-box protein (Bai et al., 1996; Feldman et al., 1997; Skowyra et al., 1997). The C-terminal β-transducin repeat domain of the F-box protein binds the target protein (Deshaies, 1999; Smith et al., 1999). In this way, F-box proteins constitute the substrate-specificity in SCF complexes.

Approximately 100 mammalian F-box proteins that form SCF complexes have been identified (Kanarek and Ben-Neriah, 2012). β-TrCP is an important member of this large family. It specifically binds the phosphorylated amino acid (aa) consensus sequence motif Asp-phosphoSer-Gly-aa-aa-phosphoSer in its target proteins (Frescas and Pagano, 2008; Fuchs et al., 2004; Yaron et al., 1997). Among its targets are IκB (Spencer et al., 1999; Suzuki et al., 1999; Tan et al., 1999; Winston et al., 1999; Yaron et al., 1998) and β-catenin (Hart et al., 1999; Jiang and Struhl, 1998; Latres et al., 1999). Thus, β-TrCP has a fundamental role in the regulation of both the canonical NF-κB and the Wnt/β-catenin signalling pathway (Fuchs et al., 2004).

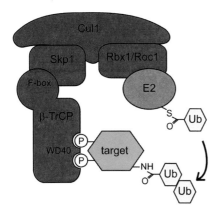

Figure 1.5: The SCF$^{\beta\text{-TrCP}}$ ubiquitin ligase complex.

The SCF$^{\beta\text{-TrCP}}$ complex (red) mediates the ubiquitin transfer from ubiquitin-conjugating enzymes (E2) onto target proteins. The β-TrCP subunit provides the target selectivity. Details on the subunits that constitute the SCF$^{\beta\text{-TrCP}}$ complex are provided in the text of Section 1.4.1. The figure is adapted from (Maniatis, 1999).

1.4.2   The paralogues β-TrCP1/FWD1 and β-TrCP2/HOS

The β-TrCP subfamily of F-box proteins include drosophila Slimb protein (Jiang and Struhl, 1998) and xenopus β-TrCP (Spevak et al., 1993) as well as the two mammalian paralogues β-TrCP1 and β-TrCP2, alternatively referred to as FWD (Margottin et al., 1998; Yaron et al., 1998) and Homologous to Slimb (HOS) (Bhatia et al., 2002; Fuchs et al., 1999), respectively.

The subfamily members are highly conserved within the functionally important F-box domain and β-transducin repeats but differ in their N-termini (Bhatia et al., 2002; Seo et al., 2009).

Generally, β-TrCP1/FWD1 and β-TrCP2/HOS are considered to be functionally redundant in the NF-κB as well as the Wnt/β-catenin signalling pathway (Frescas and Pagano, 2008). This notion of redundancy has been inferred from (i) the identical biochemical properties of the paralogues *in vitro* (Frescas and Pagano, 2008), (ii) their apparent reciprocal substitution in paralogue-specific small interfering RNA (siRNA)-mediated knock-down experiments (Guardavaccaro et al., 2003), and (iii) the general viability of β-TrCP1/FWD1 knock-out mice (Frescas and Pagano, 2008; Guardavaccaro et al., 2003; Nakayama et al., 2003).

However, β-TrCP1/FWD1 and β-TrCP2/HOS differ in several fundamental aspects. The two paralogues are encoded by two distinct genes that are differentially regulated (Chiaur et al., 2000; Koike et al., 2000). It has been shown that the expression of β-TrCP1/FWD1 as well as β-TrCP2/HOS are controlled by Wnt/β-catenin signalling (Fuchs et al., 2004). Yet, they are differentially affected by the pathway. Wnt/β-catenin signalling directly inhibits the expression and activity of β-TrCP2/HOS (Spiegelman et al., 2002b). In contrast, β-TrCP1/FWD1 is up-regulated in response to Wnt/β-catenin signalling via an indirect mechanism (Spiegelman et al., 2000). That is, its mRNA is protected from degradation by the coding region determinant-binding protein (CRD-BP), which by itself is a direct target gene product of Wnt/β-catenin signalling (Elcheva et al., 2009; Noubissi et al., 2006). Xenopus β-TrCP was also reported to be directly activated by β-catenin in complex with the xenopus TCF member XTcf-3; however, this regulatory mechanism might be restricted to the amphibian β-TrCP gene (Ballarino et al., 2004).

Besides Wnt/β-catenin signalling, stress-induced c-Jun N-terminal kinase (JNK) signalling and Akt/Protein Kinase B (PKB) signalling were reported to influence β-TrCP1/FWD1 expression (Spiegelman et al., 2000; Spiegelman et al., 2001). Furthermore, β-TrCP1/FWD1 expression was inhibited by bone morphogenetic protein 2 (BMP-2) via mitogen-activated protein kinase (MAPK) signalling (Zhang et al., 2009). In contrast, β-TrCP2/HOS expression was induced via Ras-initiated MAPK signalling (Spiegelman et al., 2002a). The control of β-TrCP1/FWD1 and β-TrCP2/HOS expression levels is considered to be a very important factor in the regulation of NF-κB and Wnt/β-catenin signalling (Bhatia et al., 2002; Fuchs et al., 2004; Kanarek and Ben-Neriah, 2012; Spiegelman et al., 2000).

β-TrCP1/FWD1 and β-TrCP2/HOS are thought to be generally low expressed and show opposite expression patterns (Fuchs et al., 2004). While β-TrCP1/FWD1 was predominantly detected in primary tumours and cancer cell lines (Fuchs et al., 2004; Saitoh and Katoh, 2001), β-TrCP2/HOS was observed in wild type tissues (Saitoh and Katoh, 2001; Spiegelman et al., 2002b). Although β-TrCP1/FWD1 up-regulation is detected in many human cancer and associated with bad survival prognosis (Frescas and Pagano, 2008), no loss of function mutations and hardly any gene alterations have been detected in the β-TrCP paralogues (Amit and Ben-Neriah, 2003; Gerstein et al., 2002; Kim et al., 2007a; Reifenberger et al., 2002; Wolter et al., 2003).

In the light of the high degree of 77% sequence homology between the β-TrCP paralogues (Fuchs et al., 1999) and due to the general notion of their functional redundancy (Frescas and Pagano, 2008), it is often not discriminated between β-TrCP1/FWD1 and β-TrCP2/HOS in publications. In order to avoid unintended confusions, the following terminology is used throughout this thesis. Whenever both paralogues are collectively addressed the term β-TrCP will be used. In any other case, the alternative names FWD1 and HOS will substitute for β-TrCP1/FWD1 and β-TrCP2/HOS, respectively.

1.5   Research objectives

To summarise Section 1.1 to Section 1.4: ubiquitination is a central process in the regulation of the canonical NF-κB and Wnt/β-catenin signalling pathway. It is for instance mediated by β-TrCP, which plays an important role in controlling the activity of both signalling pathways. The main objective of this thesis is to explore the potential impact of variations in the availability of β-TrCP on the signalling dynamics of the Wnt/β-catenin and the canonical NF-κB pathway. There exist several mechanisms to change the β-TrCP concentration that is available for Wnt/β-catenin and canonical NF-κB signalling in a cell. This thesis focusses on two specific examples: (i) concentration changes due to transcriptional regulation of β-TrCP, and (ii) alteration of β-TrCP that is available for one of the two signalling pathways due to its sequestration into the other signalling pathway.

β-TrCP abundance was reported to be transcriptionally regulated by the Wnt/β-catenin signalling pathway (Elcheva et al., 2009; Noubissi et al., 2006; Spiegelman et al., 2000; Spiegelman et al., 2002b), which directly leads to two implications. First, β-TrCP mediates transcriptional feedback mechanisms in the Wnt/β-catenin pathway (red dashed arrow I in Figure 1.6) that may modulate the dynamics of signal transduction in the pathway. In addition, transcriptional regulation of β-TrCP by β-catenin/TCF may enable the Wnt/β-catenin pathway to influence the dynamics of canonical NF-κB signal transduction (red dashed arrow II in Figure 1.6).

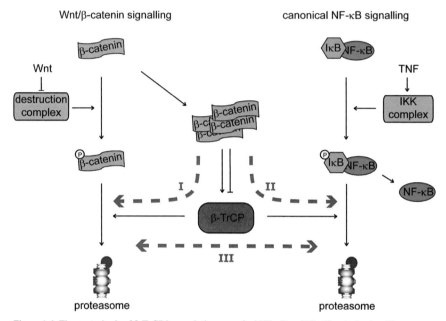

Figure 1.6: The central role of β-TrCP in regulating canonical NF-κB and Wnt/β-catenin signalling.
The scheme illustrates the central role of β-TrCP in the canonical NF-κB and Wnt/β-catenin signalling pathway. It summarises the three mechanisms of regulation of pathway activity by β-TrCP availability that are investigated in this thesis. (I) The regulation of Wnt/β-catenin signal dynamics by a positive and a negative transcriptional feedback via FWD1 and HOS, respectively. (II) The potential influence of the Wnt/β-catenin signalling pathway on canonical NF-κB signalling via transcriptional regulation of β-TrCP abundance. (III) The regulation of β-TrCP availability via its competitive sequestration into the Wnt/β-catenin and the canonical NF-κB signalling pathway.

Besides transcriptional regulation, β-TrCP availability may also vary due to sequestration effects. β-TrCP is shared between the Wnt/β-catenin and the canonical NF-κB signalling pathway (red dashed arrow III in Figure 1.6). Its competitive association with either IκB or β-catenin might result in its limited availability to one or the other pathway. In the following sections, three main questions are formulated that focus on possible implications of changes of β-TrCP availability in canonical NF-κB and Wnt/β-catenin signalling.

1.5.1 What effects may transcriptional regulation of β-TrCP abundance have on NF-κB dynamics?

The rapid response of canonical NF-κB signalling depends on the efficient degradation of IκB within minutes upon TNF stimulation (Kanarek and Ben-Neriah, 2012). Generally, the phosphorylation of IκB by IKKβ is assumed to be the rate limiting reaction in the degradation process (Kanarek and Ben-Neriah, 2012). This notion is based on biochemical studies in a cell-free system (Yaron et al., 1997). However, the rate limiting reaction might as well vary in different physiological contexts (Kanarek and Ben-Neriah, 2012). Alternative suggestions have placed it at the level of ubiquitination (Fuchs et al., 2004) or proteasome activation (Yang et al., 2001). Investigating the consequences of limited IκB ubiquitination is of special interest since novel therapeutic approaches to treat cancer, immune, and inflammatory diseases aim to directly target this process (Frankland-Searby and Bhaumik, 2012; Kanarek and Ben-Neriah, 2012). Besides pharmacological interventions, variations in the IκB ubiquitination rate could also be caused by changes in the concentration of the ubiquitin ligase subunit β-TrCP under physiological conditions. Indeed, it has been shown that Wnt/β-catenin signalling can affect canonical NF-κB activation through the regulation of β-TrCP abundance (Spiegelman et al., 2000; Spiegelman et al., 2002b).

Chapter 3 focuses on the possible impact of variations in β-TrCP concentrations on canonical NF-κB dynamics. To investigate this question, a detailed kinetic model of canonical NF-κB signalling (Lipniacki et al., 2004) is modified and analysed. The modelling approach additionally reveals the potential impact of the Wnt/β-catenin pathway on canonical NF-κB signalling by means of changing β-TrCP concentration due to transcriptional regulation of β-TrCP expression.

1.5.2 Are the two paralogues of β-TrCP functionally redundant in the Wnt/β-catenin signalling pathway?

The roles of the two paralogues of β-TrCP, FWD1 and HOS, in the Wnt/β-catenin signalling pathway still pose unsolved questions regarding fundamental aspects (Lau et al., 2012). One aspect concerns the apparently redundant role of both paralogues in β-catenin ubiquitination. It is still unknown what functional advantage this redundancy would offer (Kanarek and Ben-Neriah, 2012).

A related question deals with the functional relevance of the opposite regulation of the two paralogues by β-catenin/TCF (Amit and Ben-Neriah, 2003; Kanarek and Ben-Neriah, 2012). It is generally expected that the opposite regulation mechanisms may offset each other, that is cancel their individual impact, unless the paralogues would differ in their substrate specificity. In this respect, it is important to note that these different mechanisms establish a positive (FWD1) and a negative feedback (HOS) in the Wnt/β-catenin signalling pathway. These two feedbacks may allow for distinct roles of FWD1 and HOS in Wnt/β-catenin signalling, contrasting the functional redundancy discussed above.

In Chapter 4, the impacts of the feedbacks mediated by FWD1 and HOS on the dynamics of Wnt/β-catenin signalling are investigated. To this end, a detailed model of the Wnt/β-catenin signalling pathway (Lee et al., 2003) is extended to include these two feedback mechanisms.

1.5.3 Under what conditions would crosstalk via competitive β-TrCP sequestration yield observable effects on β-catenin and/or NF-κB dynamics?

β-TrCP is involved in the ubiquitination of many different proteins (Frescas and Pagano, 2008; Fuchs et al., 2004). A theoretical advantage of the involvement of β-TrCP in multiple pathways is to facilitate the coordination of their respective activities (McNeill and Woodgett, 2010). Since β-TrCP is thought to be low expressed it has been suggested that its different target proteins may compete for this limited pool of β-TrCP and thereby influence the degradation of each other (Fuchs et al., 2004). Indeed, this concept seems to be supported by several experimental reports. For instance, the viral protein unique (Vpu) of human immunodeficiency virus 1 (HIV-1) is a direct target protein of β-TrCP that assists in the proteasomal degradation of cluster of differentiation 4 (CD4) in HIV-infected T helper cells. Vpu has been shown to interfere with the degradation of IκBα by acting as a competitive inhibitor of the β-TrCP-IκBα-interaction (Besnard-Guerin et al., 2004; Bour et al., 2001). It has furthermore been reported that overexpression of activating transcription factor (ATF), a further target protein of β-TrCP, results in an accumulation of β-catenin (Besnard-Guerin et al., 2004). ATF interferes with the constitutive degradation of β-catenin mediated by β-TrCP, although it is still unresolved whether ATF acts via a competitive mechanism of inhibition.

β-TrCP was shown to recognise the identical phosphorylated amino acid sequences in IκB and in β-catenin. Consequently, one particular $SCF^{\beta\text{-}TrCP}$ ubiquitin ligase complex is sequestered for either IκB or β-catenin degradation (Ougolkov et al., 2004; Spiegelman et al., 2000;

Winston et al., 1999). This competitive sequestration is a proposed mechanism for β-catenin-induced activation of NF-κB that has been experimentally observed in human embryonic kidney 293T cells and mouse embryonic fibroblast NIH3T3 cells (Fuchs et al., 2004; Spiegelman et al., 2000; Spiegelman et al., 2002b).

In Chapter 5, a minimal model of canonical NF-κB and Wnt/β-catenin signalling linked through competitive β-TrCP binding is developed. This minimal model is used to investigate under what conditions both pathways may influence the dynamics of each other via competitive β-TrCP sequestration.

2. Mathematical modelling approaches

Signalling pathways are regulated through complex interactions between many signalling molecules forming signalling networks. The complex nature of these signalling networks renders it very difficult to understand how a signal propagates through the network and which parameters determine the biologically relevant functional outcome. Mathematical modelling using ordinary differential equations (ODEs) provides a method to gain insight into the mechanisms that determine signalling dynamics (Klipp and Liebermeister, 2006; Tyson et al., 2003; Wolkenhauer et al., 2005). In this modelling approach, the ODEs describe the dynamical behaviour of the concentrations of each species in the model over time.

Several ODE models of Wnt/β-catenin signalling and NF-κB signalling have been developed in the last decade, reviewed in (Kofahl and Wolf, 2010) and (Basak et al., 2012; Cheong et al., 2008), respectively. The models range from detailed models that describe molecular interactions of pathway components (Cho et al., 2003; Hoffmann et al., 2002; Kim et al., 2009; Lee et al., 2003; Schmitz et al., 2013; Schmitz et al., 2011) to very abstract models, which focus on general properties of the pathway dynamics (Krishna et al., 2006; Mirams et al., 2010). The scopes of the various models cover a multitude of different biological topics. In the case of Wnt/β-catenin signalling, mathematical models have for instance been used to exploit the interrelation between adhesive and transcriptional functions of β-catenin as well as the consequences of carcinogenic mutations of pathway components (Benary et al., 2013; Cho et al., 2006; van Leeuwen et al., 2007). The analyses of NF-κB signalling models have initiated a discussion on the potential relevance of oscillations of nuclear NF-κB concentration in the regulation of target gene expression (Barken et al., 2005; Nelson et al., 2005; Nelson et al., 2004; Nikolov et al., 2009). NF-κB signalling models have also been used to investigate how the canonical and non-canonical NF-κB pathway branches may influence each other (Basak et al., 2012).

Two detailed kinetic models, one describing Wnt/β-catenin signalling (Lee et al., 2003) and one describing canonical NF-κB signalling (Lipniacki et al., 2004), provide the starting point of this study. The two detailed kinetic models were chosen because they have been well validated by experiments. Model parameters have been either measured or good estimates have been derived to quantitatively describe the temporal concentration changes of the components in the signalling pathways as observed in experiments (Lee et al., 2003; Lipniacki et al., 2004). The detailed kinetic models are introduced in Section 2.1 and Section 2.2. In Section 2.3, alternative parameter sets of the Wnt/β-catenin signalling model are specified that have been published to simulate carcinogen mutations of pathway components (Cho et al., 2006). In the two detailed kinetic models, β-TrCP is not explicitly considered. Therefore, the models are modified and extended to address the questions of Section 1.5. The actual changes in the models are specified in Section 3.1, Section 4.1, and Section 5.1.

In Section 2.4 to Section 2.8, mathematical tools to study dynamical systems are introduced. In Section 2.4 and Section 2.5, quantitative measures are defined that characterise dynamical properties of signal responses and crosstalk in transient dynamics, respectively. In Section 2.6 and Section 2.7, sensitivity coefficients are defined and bifurcation analysis is introduced. Both approaches are used to explore the dynamical behaviour of the models. Finally, Kendall rank correlation, which is a statistical method used in this study, is introduced in Section 2.8.

2.1 The detailed kinetic model of canonical NF-κB signalling

The detailed kinetic model of canonical NF-κB signalling (Lipniacki et al., 2004) quantitatively describes the molecular processes that transduce an extracellular TNF signal into a change of nuclear NF-κB concentration. The model accounts for transient TNF-dependent IKK activation, NF-κB-regulated target gene expression, and the inhibitory action of IκB and A20 on NF-κB activation (Figure 2.1).

In the model, three types of the IKK complex are taken into account: a neutral form (IKKn), an active form (IKKa), and an inactive form (IKKi). All three types are degraded (Reaction 2, Reaction 5, and Reaction 6 in Figure 2.1), but only IKKn is produced *de novo* (Reaction 1). IKKn is activated by TNF stimulation (Reaction 3). TNF is implemented into the model as a logical variable that is either set to 0 in the absence of a stimulus, or to 1 if a stimulus is present (Equation 7.44, Appendix 7.1). Besides IKKn activation, TNF also promotes the inactivation of IKKa (Reaction 26). In addition, IKKa inactivates itself in a TNF-independent manner (Reaction 4). IKKa binds NF-κB-bound and unbound IκB (Reaction 7 and Reaction 9, respectively). IκB is degraded in an IKKa-dependent (Reaction 8 and Reaction 10) and IKKa-independent manner (Reaction 15 and Reaction 21). The IKKa-dependent degradation of IκB via Reaction 10 results in the dissociation of the IKK/IκB/NF-κB complex liberating NF-κB from its inhibitor. NF-κB can then translocate into the nucleus (Reaction 11).

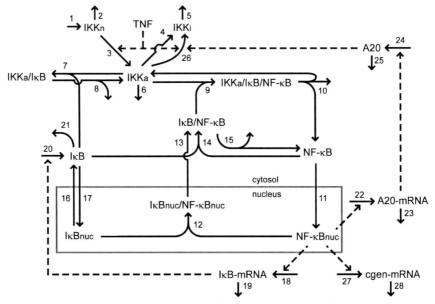

Figure 2.1: Reaction scheme underlying the detailed kinetic model of canonical NF-κB signalling.
The schematic representation of the detailed kinetic model of canonical NF-κB signalling (Lipniacki et al., 2004) is shown. In the scheme, the numbers next to the arrows specify the number of the reaction. Reaction rates and parameters of the model are given in Appendix 7.1. One-headed arrows denote reactions taking place in the indicated direction. Dashed arrows illustrate activations. Components in a complex are separated by slashes. The grey box surrounds the species in the nuclear compartment of the model. Additional explanations are provided in the text of Section 2.1. cgen: control gene; nuc: nuclear.

In the nucleus, NF-κB regulates the transcription of IκB-mRNA, A20-mRNA, and the control gene transcript cgen-mRNA (Reaction 18, Reaction 22, and Reaction 27, respectively). IκB-mRNA, A20-mRNA, and cgen-mRNA are degraded via Reaction 19, Reaction 23, and

Reaction 28, respectively. IκB-mRNA and A20-mRNA are furthermore translated into their respective proteins (Reaction 20 and Reaction 24, respectively). A20 protein promotes the inactivation of IKKa (Reaction 26) and is degraded via Reaction 25. IκB can either associate with NF-κB in the cytoplasm to form IκB/NF-κB complexes (Reaction 14) or shuttle between the cytosolic and nuclear compartment (Reaction 16 and Reaction 17). Nuclear IκB (IκBnuc) associates with nuclear NF-κB (NF-κBnuc) to form nuclear IκBnuc/NF-κBnuc complexes (Reaction 12). These nuclear complexes translocate from the nuclear to the cytosolic compartment (Reaction 13).

Overall, the detailed kinetic model of canonical NF-κB consists of 14 ODEs and one conservation relation for NF-κB (see Appendix 7.1). The original parameters are listed in Table 7.1 (Appendix 7.1). The detailed kinetic model is used in Chapter 3 to analyse the impact of changes in the rate constants of both IKK-dependent IκB degradation reactions (Reaction 8 and Reaction 10) on the dynamical properties of the nuclear NF-κB response to TNF stimulation. In Chapter 5, the detailed kinetic model constitutes the starting point to develop a reduced canonical NF-κB signalling sub-module in the minimal model of competitive β-TrCP sequestration.

2.2   The detailed model of Wnt/β-catenin signalling

The detailed model of Wnt/β-catenin signalling (Lee et al., 2003) consists of 15 ODEs (Equations 7.45 - 7.59, Appendix 7.2). This set of differential equations has been simplified to eight algebraic and seven differential equations (Lee et al., 2003) taking into account conservation relations and rapid equilibrium approximations of binding reactions (Section 7.2.2 and Section 7.2.3, respectively). The original parameters are listed in Table 7.3, Table 7.4, and Table 7.5 in Appendix 7.2. A schematic representation of the detailed model of Wnt/β-catenin signalling is shown in Figure 2.2.

In the model, β-catenin is continuously produced *de novo* (Reaction 12). It can reversibly form complexes with APC, TCF, and the destruction complex APC*/Axin*/GSK3 (Reaction 17, 16, and 8, respectively). If bound in the APC*/Axin*/GSK3/β-catenin complex, β-catenin is phosphorylated (Reaction 9) and subsequently released from the complex (Reaction 10) to be degraded via Reaction 11. The destruction-complex-mediated degradation of β-catenin is inhibited by Wnt stimulation. A Wnt stimulus activates Dsh (Reaction 1) and activated Dsh (Dsh$_a$) promotes the dissociation of GSK3 from the APC/Axin/GSK3 complex (Reaction 3). Dsh$_a$ is inactivated via Reaction 2. In addition to the destruction-complex-mediated degradation, β-catenin can also be degraded in an alternative way (Reaction 13).

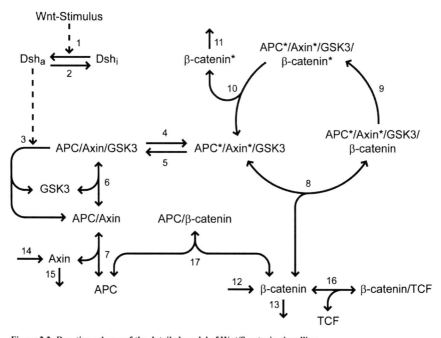

Figure 2.2: Reaction scheme of the detailed model of Wnt/β-catenin signalling.
A schematic representation of the detailed model of Wnt/β-catenin signalling (Lee et al., 2003) is shown. In the scheme, the number next to the arrows specifies the number of the reaction. Reaction rates and parameters of the model are given in Appendix 7.2. One-headed arrows denote reactions taking place in the indicated direction. Double-headed arrows represent binding reactions. Dashed arrows illustrate activations. The asterisk marks phosphorylation of the component. Components in a complex are separated by a slash.

Similar to β-catenin, Axin is produced *de novo* (Reaction 14). Axin is degraded via Reaction 15. It reversibly binds APC to form APC/Axin complexes (Reaction 7), which then reversely bind GSK3 (Reaction 6) to form APC/Axin/GSK3 complexes. In the APC/Axin/GSK3 complex, APC and Axin can be phosphorylated and dephosphorylated (Reaction 4 and Reaction 5, respectively).

In Chapter 4, the detailed model is modified and extended by two transcriptional feedback mechanisms acting via the β-TrCP paralogues HOS and FWD1. The resulting extended model is referred to as two-feedback model (Section 4.1). In Chapter 5, the detailed model of Wnt/β-catenin signalling constitutes the starting point to develop a reduced Wnt/β-catenin sub-module in the minimal model of competitive β-TrCP sequestration.

2.3   Simulation of Wnt/β-catenin signalling in cancer cells

Mutations of components of the Wnt/β-catenin pathway often result in an aberrant activation of Wnt/β-catenin signalling. For instance, mutations in APC have been detected in more than 85% of colon cancer samples (Giles et al., 2003). Most of these mutations lead to truncations in the APC protein such that different number of Axin and/or β-catenin binding sites are missing in the mutant APC. To model this impaired complex formation of truncated versions of APC with Axin and/or β-catenin, additional parameter sets of 13 different APC mutants (referred to as "m1" to "m13") for the detailed model of Wnt/β-catenin signalling have been published (Cho et al., 2006). These parameter sets differ in only three parameters that relate to (i) the dissociation of the APC/β-catenin complex (Reaction 17 in Figure 2.2), (ii) the dissociation of the APC/Axin complex (Reaction 7), and (iii) the dissociation of β-catenin from the APC*/Axin*/GSK3/β-catenin complex (Reaction 8). In this thesis, these parameter sets (APC mutants "m1" to "m13" listed in Table 7.9 in Appendix 7.3.5) are adapted in order to simulate APC mutant cancer cells in the context of the two-feedback model, which is described in Section 4.1.

2.4   Definition of signalling time, signal duration, and signal amplitude

In this thesis, ODEs are used to describe the continuous dynamics of the concentrations of the species in the signalling pathways. The change in the concentrations of a species $C[t]$ over time is given by a sum of rate equations. A rate equation consists of rate laws, which describe the biochemical kinetics of the reactions in the model. A positive rate equation indicates producing reactions resulting in an increase in the concentration of the species. A negative rate equation indicates consuming reactions leading to a decrease in the concentration of the species. If for each model species the rate equations of producing and consuming reactions sum-up to zero the concentrations of all model species remain constant over time. This condition is referred to as steady state of the model. Steady states are calculated by setting the time derivative of each species concentration in the model to zero and solving the resulting algebraic equation system for the species concentrations. They are used in this thesis as initial conditions to simulate the continuous dynamics of the model species over time.

To quantitatively characterise the dynamics of species in ODE models, several measures have been introduced (reviewed in (Klipp, 2009)). Here, three particular measures are considered: signalling time, signal duration, and signal amplitude (Heinrich et al., 2002; Kofahl and Klipp, 2004; Llorens et al., 1999). These three measures may be interpreted to provide information on (i) the expected time that the stimulus signal needs to arrive at the level of pathway readout (signalling time), (ii) the duration of the response at the level of pathway readout (signal duration), and (iii) the magnitude of this response (signal amplitude) (Heinrich et al., 2002). In this thesis, the following definitions of signalling time (Equation 2.1), signal duration (Equation 2.2), and signal amplitude (Equation 2.3) are used:

$$\text{signalling time} = \frac{\int_{t_{\text{initial}}}^{t_{\text{final}}} t \cdot \left|\frac{dC[t]}{dt}\right| dt}{\int_{t_{\text{initial}}}^{t_{\text{final}}} \left|\frac{dC[t]}{dt}\right| dt} \qquad 2.1$$

$$\text{signal duration} = \sqrt{\frac{\int_{t_{\text{initial}}}^{t_{\text{final}}} t^2 \cdot \left|\frac{dC[t]}{dt}\right| dt}{\int_{t_{\text{initial}}}^{t_{\text{final}}} \left|\frac{dC[t]}{dt}\right| dt} - \left(\frac{\int_{t_{\text{initial}}}^{t_{\text{final}}} t \cdot \left|\frac{dC[t]}{dt}\right| dt}{\int_{t_{\text{initial}}}^{t_{\text{final}}} \left|\frac{dC[t]}{dt}\right| dt}\right)^2} \qquad 2.2$$

$$\text{signal amplitude} = \max C[t] - \min C[t] \qquad t_{\text{initial}} \leq t \leq t_{\text{final}} \qquad 2.3$$

These definitions can be applied to the wide range of different dynamical responses displayed by the mathematical models in this thesis. They do not require that the signal eventually returns to its initial steady state and are also suitable for damped oscillatory transitions. In addition, the definitions can be adjusted to deal with sustained oscillations (Llorens et al., 1999), although, this extension is not considered in this thesis.

2.5 Definition of a measure to quantify crosstalk impact in transient dynamics

In a strong, but often used, simplification, a signal transduction pathway may be considered to constitute a route from one pathway-specific stimulus input (e.g., TNF or Wnt) to a component considered as pathway-specific output (e.g., NF-κB and β-catenin, respectively). However, very often signalling pathways share one or more components. In this case, mutual interaction and/or interference between the signalling pathways may take place. The signal, which is considered specific for one pathway, may pass through the shared component and influence the output that is considered specific for the other pathway. The other pathway is thus affected by a pathway-extrinsic stimulus. This phenomenon is referred to as crosstalk.

To quantify crosstalk, different measures have been introduced such as pathway specificity and fidelity (Bardwell et al., 2007; Komarova et al., 2005; Schaber et al., 2006). Briefly, pathway specificity characterises signalling networks that allow one stimulus to specifically affect only one output, while pathway fidelity relates to networks in which a particular output is merely regulated by one stimulus. Chapter 5 explores how strong a pathway-extrinsic stimulus influences the pathway-intrinsic output using the example of the minimal model of competitive β-TrCP sequestration. There, the impact of simultaneous TNF stimulation on the β-catenin dynamics in response to Wnt stimulation is investigated as well as the impact of simultaneous Wnt stimulation on the NF-κB dynamics in response to TNF stimulation. To quantify the effects, the measure of crosstalk impact (CI) is introduced:

$$CI = \frac{\max \left| C[t]_{single} - C[t]_{simultaneous} \right|}{\text{total concentration range}} \qquad 2.4$$

$$\text{total concentration range} = \max \left\{ C[t]_{single}, C[t]_{simultaneous} \right\} - \min \left\{ C[t]_{single}, C[t]_{simultaneous} \right\} \qquad 2.5$$

Where $C[t]_{single}$ represents the dynamics of the pathway-intrinsic output to only the pathway-specific stimulus, while $C[t]_{simultaneous}$ represents the dynamics of the pathway-intrinsic output when both stimuli (pathway-specific and pathway-extrinsic) are applied simultaneously. The total concentration range (Equation 2.5) of the pathway-intrinsic output dynamics in the simulations is given by the difference between the overall maximal and minimal value of $C[t]_{single}$ and $C[t]_{simultaneous}$. A $CI \geq 0.05$ is considered to indicate visually distinct output dynamics upon either simultaneous stimulation with TNF as well as Wnt or stimulation with the pathway-specific stimulus alone at a certain time point t in the simulations. In contrast, a CI < 0.05 is considered to yield no observable modulation of the pathway-specific output by the pathway-extrinsic stimulus.

2.6   Calculation of sensitivity coefficients

A sensitivity analysis indicates how robust certain dynamical properties are with respect to the choice of the values of model parameters. The approach is originally based on the theoretical framework of metabolic control theory (Heinrich and Rapoport, 1974; Kacser and Burns, 1973; Reder, 1988). However, it has also been successfully applied to study signalling models (Hu and Yuan, 2006; Kholodenko et al., 1997; Kruger and Heinrich, 2004; Yi et al., 2007). The sensitivity analyses performed in this thesis focus on the robustness of steady state concentrations. In the analysis each parameter, one at a time, is increased and decreased by 0.1% of its nominal value. In this case, the sensitivity coefficient of steady state concentration S with respect to a parameter p is given by

$$\text{sensitivity coefficient} = \frac{S[p^+] - S[p^-]}{0.002 \cdot S[p]} \qquad\qquad 2.6$$

where $S[p^+]$, $S[p^-]$, and $S[p]$ refer to the calculated steady state concentrations for 0.1% increased, 0.1% decreased, and the unchanged nominal value of parameter p, respectively. Since the perturbation of the parameter p is so small, the following three statements concerning the sensitivity coefficient hold in this thesis: (i) a positive sensitivity coefficient indicates that an increase of the value of parameter p leads to an increase of the steady state concentration S, (ii) a negative sensitivity coefficient indicates that the increase of the parameter value results in a decrease of the steady state concentration, and (iii) for positive and negative sensitivity coefficient holds that the larger the absolute value of the sensitivity coefficient the larger is the change of the steady state concentration in response to the parameter perturbation.

2.7   Bifurcation analysis

A bifurcation is a qualitative change in the behaviour of a (non-linear) ODE system due to a small change in one parameter, the bifurcation parameter. In the present case, a change of the number of steady states and/or a change of the steady state stability in response to small perturbations of the bifurcation parameter is considered. There exist several open source software tools for bifurcation analysis, such as MATCONT[2] and XPP-AUT[3]. Unfortunately, both tools could not perform the analysis of the models used in this thesis in a stable and reliable way. Using an alternative approach in this thesis, bifurcation analysis is performed by numerically calculating all steady states and their stability for distinct values of the bifurcation parameter using Mathematica 8.0 (Wolfram Research). To that end, steady states are calculated by setting the time derivative of each species concentration in the model to zero and solving the resulting algebraic equation system for the species concentrations. The number of steady states is calculated by counting the steady state solutions that yield non-negative real values of all species concentrations. To determine the stability of a steady state, the eigenvalues of the Jacobian matrix are calculated (Heinrich and Schuster, 1996). The Jacobian matrix is the matrix of all first-order partial derivatives of the ODEs with respect to the species of the model. If the real parts of all eigenvalues are negative the steady state is asymptotically stable. Asymptotically stable means that after a small initial perturbation of the steady state the model species will return to the original steady state again. Otherwise, the steady state is called unstable. A special example of a bifurcation, which is related to an interconversion of a stable and an unstable steady state due to small perturbations of the bifurcation parameter, is the Hopf bifurcation. At Hopf bifurcations, limit cycle oscillations arise when the steady state becomes unstable. The Hopf bifurcation is detected if all eigenvalues of the Jacobian matrix have negative real parts with the exception of one conjugate non-zero purely imaginary pair.

---

[2] MatCont Dhooge, A., Govaerts, W., and Kuznetsov, Y.A. (2003). MATCONT: A MATLAB package for numerical bifurcation analysis of ODEs. ACM Trans Math Softw *29*, 141-164. is a MATLAB software project developed under the supervision of W. Govaerts (Gent, Belgium) and Yu. A. Kuznetsov (Utrecht, The Netherlands) and is freely available at: http://sourceforge.net/projects/matcont

[3] XPPAUT is created by Bard Ermentrout (Pittsburgh, Pennsylvania, USA) and is distributed under the GNU public license: http://www.math.pitt.edu/~bard/xpp/xpp.html

Bifurcations can be visualised in bifurcation diagrams. In these diagrams, usually the steady states of the system are plotted versus the bifurcation parameter. In this thesis, stable steady states are indicated by solid lines in bifurcation diagrams, while unstable steady states are visualised by dashed lines. In addition, the maximal value and minimal value of the amplitude of limit cycle oscillations is marked by dotted lines in the bifurcation diagrams.

2.8   Kendall rank correlation

In Chapter 5, correlations of the values of parameters and parameter combinations with respect to steady state concentrations and crosstalk impact (Section 2.5) are investigated. To that end, Kendall rank correlation coefficients are calculated (Kendall, 1938). The Kendall rank correlation coefficient has been used as a non-parametric measure of association between two sets of observations X and Y on the basis of their ranking. Kendall rank correlation coefficient between X and Y is given by Equation 2.7:

$$\text{Kendall rank correlation coefficient} = \frac{n_c - n_d}{\sqrt{(n_c + n_d + n_x) \cdot (n_c + n_d + n_y)}} \qquad 2.7$$

where n_c is the number of concordant pairs of observations, n_d is the number of discordant pairs, n_x is the number of ties involving only the X variable, and n_y is the number of ties involving only the Y variable. A pair of observations $\{X_i, Y_i\}$ and $\{X_j, Y_j\}$ are said to be concordant if the ranks for both elements agree. That is, if both $X_i > X_j$ and $Y_i > Y_j$ or if both $X_i < X_j$ and $Y_i < Y_j$. A pair of observations $\{X_i, Y_i\}$ and $\{X_j, Y_j\}$ are said to be discordant, if $X_i > X_j$ and $Y_i < Y_j$ or if $X_i < X_j$ and $Y_i > Y_j$. If $X_i = X_j$ or $Y_i = Y_j$, the pair is considered to be tied. The Kendall rank correlation coefficient takes values in the range from -1 to 1; with 1 corresponding to perfect agreement in the two rankings, -1 corresponding to perfect disagreement in the two rankings. If both sets are independent of each other, the Kendall rank correlation coefficient is zero. To calculate the Kendall rank correlation coefficients and their respective p-values (under the null hypothesis of Kendall rank correlation coefficient = 0) MATLAB 2010b (MathWorks) is used. The smaller the p-value the less likely is the observed rank correlation as a result of random chance under the null hypothesis.

3. Investigation of the impact of transcriptional regulation of β-TrCP abundance on canonical NF-κB signalling

In this chapter, the potential impact of variations in β-TrCP abundance on the dynamics of nuclear NF-κB in response to TNF stimulation is investigated. Starting point of the investigation is a detailed kinetic model of canonical NF-κB signalling (Lipniacki et al., 2004) introduced in Section 2.1. Its analysis leads to the prediction that β-TrCP abundance influences steady state concentration and steady state stability of nuclear NF-κB as well as characteristic properties of the transient dynamics of nuclear NF-κB such as its amplitude.

3.1 Modification of the detailed model of canonical NF-κB signalling

To investigate the impact of transcriptional regulation of β-TrCP abundance on canonical NF-κB signalling, the detailed kinetic model of canonical NF-κB signalling, introduced in Section 2.1, is slightly modified in its parameterisation. In particular, the two IKK-dependent IκB degradation rate constants k_8 and k_{10} (Equation 7.23 and Equation 7.25, respectively, in Appendix 7.1) are substituted by a single parameter $k_{\beta\text{-TrCP}}$. This approach is based on the following line of arguments. Reaction 8 and Reaction 10 (highlighted in red in Figure 3.1) describe the IKK-dependent IκB degradation process, which gathers IκB phosphorylation, IκB ubiquitination, and proteasomal degradation of IκB into a single mechanistic step in the model. Experimental observations suggest that IκB degradation is influenced by changes in the concentration of β-TrCP. In particular, overexpression of β-TrCP reduces IκB concentration by enhancing its proteasomal degradation (Bhatia et al., 2002; Fuchs et al., 1999; Kroll et al., 1999; Spencer et al., 1999; Wang et al., 2004). In contrast, overexpression of a dominant negative mutant of β-TrCP inhibited the proteasomal degradation of IκB (Bhatia et al., 2002; Fuchs et al., 1999; Kroll et al., 1999; Spencer et al., 1999; Wang et al., 2004). In the modelling approach, changes of β-TrCP concentration are presumed to modulate the rate constants k_8 and k_{10} of both IκB degradation reactions (Reaction 8 and Reaction 10 in Figure 3.1) to equal extend. As both rate constants k_8 and k_{10} have originally been set to identical values (Table 7.1, Appendix 0), both parameters can be substituted by the same parameter $k_{\beta\text{-TrCP}}$ without affecting the original dynamics of the model. The parameter $k_{\beta\text{-TrCP}}$ is set to 6 min^{-1}, which is the published original value for both rate constants k_8 and k_{10} (Lipniacki et al., 2004). Taken together, variations in β-TrCP abundance are assumed to change the rate constant of the IKK-dependent IκB degradation processes ($k_{\beta\text{-TrCP}}$) in the detailed kinetic model of canonical NF-κB signalling.

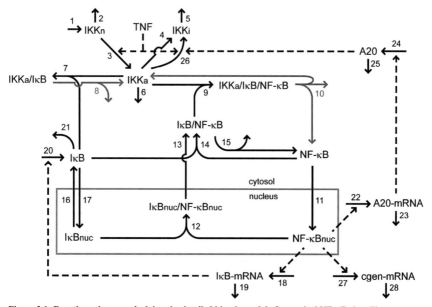

Figure 3.1: Reaction scheme underlying the detailed kinetic model of canonical NF-κB signalling.
Schematic representation of the detailed kinetic model of canonical NF-κB signalling introduced in Section 2.1 (Lipniacki et al., 2004). In the scheme, the number next to an arrow specifies the number of the reaction. One-headed arrows denote reactions taking place in the indicated direction. Dashed arrows illustrate activations. Components in a complex are separated by slashes. Reactions involving β-TrCP-mediated IκB degradation are highlighted in red. The nuclear species of the model are surrounded by the grey box. Additional explanations are provided in the text of Section 2.1. cgen: control gene transcript; nuc: nuclear.

3.2   Analysis of the model of canonical NF-κB signalling

3.2.1   A reduction of the value of $k_{\beta\text{-TrCP}}$ changes sensitivity coefficients of nuclear NF-κB

In the detailed kinetic model of canonical NF-κB signalling (Section 2.1), IKK-dependent degradation of IκB has been assumed to be limited by the complex formation of IKK/IκB and IKK/IκB/NF-κB rather than the phosphorylation, ubiquitination, and/or proteasomal degradation of IκB. The rate constants k_8 and k_{10} have therefore been set to a sufficiently large value (6 min^{-1}) in the original model (Lipniacki et al., 2004). In this chapter, however, the consequences of limited ubiquitination, due to changed β-TrCP abundance, are considered. The first step in the investigation is to determine how strong the influence of each model parameter on NF-κB dynamics changes if $k_{\beta\text{-TrCP}}$ is reduced. To this end, a sensitivity analysis, introduced in Section 2.5, is performed. The computed sensitivity coefficients (Equation 2.6) quantify the impact that subtle changes in each single parameter value have on the stimulated steady state concentration of nuclear NF-κB (NF-κBnuc). The sensitivity coefficients are calculated for two different values of $k_{\beta\text{-TrCP}}$. In the reference case, $k_{\beta\text{-TrCP}}$ is set to the published value of 6 min^{-1} (Figure 3.2, black bars). In the second case, $k_{\beta\text{-TrCP}}$ will be reduced by more than one order of magnitude to investigate the effects of strong reduction. It has been observed that in the case of a reduction of $k_{\beta\text{-TrCP}}$ by two orders of magnitude, the steady state of nuclear NF-κB becomes instable (see Section 3.2.2). Thus, the reduced value of $k_{\beta\text{-TrCP}}$ is set to 0.12 min^{-1} (Figure 3.2, grey bars), which is 1/50th of the reference value.

The sensitivity analysis considering the reference parameter value of 6 min^{-1} (Figure 3.2, black bars) shows that parameters associated with IKK production (v_1), IKK binding to IκB/NF-κB (k_9), as well as parameters involved in the IκB feedback (k_{18} to k_{20}) and A20 feedback (k_{22} to k_{26}) strongly impact the stimulated steady state concentration of nuclear NF-κB. A positive sensitivity coefficient indicates that an increase of the value of the parameter (v_1, k_9, k_{19}, k_{23}, and k_{25}) leads to an increase of the stimulated steady state concentration of nuclear NF-κB, while a negative sensitivity coefficient indicates that an increase of the parameter value (k_{18}, k_{20}, k_{22}, k_{24}, and k_{26}) results in a decrease of the steady state concentration. In contrast to the strong impact of the parameters v_1, k_9, k_{18} to k_{20}, and k_{22} to k_{26}, small perturbations in the value of $k_{\beta\text{-TrCP}}$ hardly influence the stimulated steady state concentration (Figure 3.2, black bars). This was expected for the reference value of $k_{\beta\text{-TrCP}}$

(6 min^{-1}) because the rate constants of IκB ubiquitination were originally set to be sufficiently large to have no influence (Lipniacki et al., 2004). If, however, k$_{\beta\text{-TrCP}}$ is reduced to 0.12 min^{-1} (Figure 3.2, grey bars), its influence on the stimulated steady state concentration of nuclear NF-κB increases substantially by approximately 35-fold. In contrast to the gain in sensitivity of k$_{\beta\text{-TrCP}}$, the parameters associated with IKK production (v$_1$), IKK binding to IκB/NF-κB (k$_9$), and A20 feedback (k$_{22}$ to k$_{26}$) show diminished sensitivity coefficients. They are reduced to about two thirds of their values if k$_{\beta\text{-TrCP}}$ equals 6 min^{-1} (Figure 3.2, grey bars compared to black bars). In contrast to the A20 feedback, the influence of the parameters involved in the IκB feedback (k$_{18}$ to k$_{20}$) hardly changes.

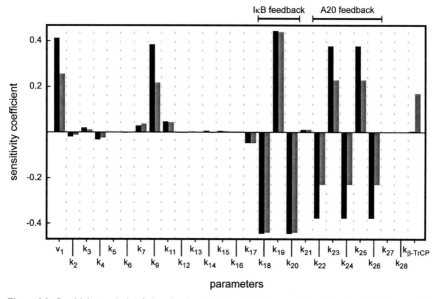

Figure 3.2: Sensitivity analysis of the stimulated steady state concentration of nuclear NF-κB towards small parameter changes for two different values of k$_{\beta\text{-TrCP}}$.

Sensitivity coefficients of each rate constant with respect to the stimulated steady state concentration of nuclear NF-κB (NF-κBnuc) are calculated (see Section 2.5) for two different values of k$_{\beta\text{-TrCP}}$. Black bars denote the reference value of k$_{\beta\text{-TrCP}}$ (6 min^{-1}) while grey bars denote a reduced value of k$_{\beta\text{-TrCP}}$ (0.12 min^{-1}). The reduction of k$_{\beta\text{-TrCP}}$ from 6 min^{-1} to 0.12 min^{-1} increases the influence of k$_{\beta\text{-TrCP}}$ on the stimulated steady state concentration of nuclear NF-κB by approximately 35-fold. The sensitivity coefficients of several other parameters are also affected. The parameters related to the IκB and A20 feedbacks are indicated as a matter of visual clarity.

For both values of $k_{\beta\text{-TrCP}}$ (6 min^{-1} and 0.12 min^{-1}), the calculated sensitivity coefficients of the parameters related to the control-gene-mRNA (k_{27} and k_{28}) are zero (Figure 3.2). This was expected because the cgen-mRNA does not influence any species in the detailed kinetic model of canonical NF-κB signalling (Figure 3.1). The calculated sensitivity coefficients of parameters related to nuclear export reactions (k_{13} and k_{16}), nuclear and cytosolic IκB/NF-κB complex formation (k_{12} and k_{14}, respectively), IκB/NF-κB-associated IκB degradation (k_{15}), and degradation of IKKi and IKKa (k_5 and k_6, respectively) are less than 0.01. Changes in these parameters have hardly an impact on the stimulated steady state concentration of nuclear NF-κB.

Taken together, a reduction of the value of $k_{\beta\text{-TrCP}}$ reduces the sensitivity coefficients of the stimulated steady state concentration of nuclear NF-κB with respect to the parameters associated with IKK production (v_1), IKK binding to IκB/NF-κB (k_9), and A20 feedback (k_{22} to k_{26}) but hardly those of the parameters involved in the IκB feedback (k_{18} to k_{20}). In addition, the influence of $k_{\beta\text{-TrCP}}$ on the stimulated steady state concentration of nuclear NF-κB increases substantially by approximately 35-fold, if $k_{\beta\text{-TrCP}}$ is reduced from 6 min^{-1} to 0.12 min^{-1}.

3.2.2 The value of $k_{\beta\text{-TrCP}}$ influences the steady state of nuclear NF-κB upon TNF stimulation

To investigate the effects of the β-TrCP concentration on the long-term response of nuclear NF-κB upon a TNF stimulus, stimulated steady state concentrations of nuclear NF-κB (NF-κBnuc) are calculated for different values of $k_{\beta\text{-TrCP}}$ between $2 \cdot 10^{-5}$ min^{-1} and 24 min^{-1}. The value of $k_{\beta\text{-TrCP}}$ is varied over such a wide range in order to cover all possible conditions, since absolute concentrations of β-TrCP that may restrict the values of $k_{\beta\text{-TrCP}}$ have not been published. The analysis shows that $k_{\beta\text{-TrCP}}$ values greater than 1 min^{-1} hardly affect the stimulated steady state concentration of nuclear NF-κB of about 16.7 nM (Figure 3.3A). Decreasing values of $k_{\beta\text{-TrCP}}$ reduce the stimulated steady state concentration of nuclear NF-κB. For values of $k_{\beta\text{-TrCP}}$ smaller than 10^{-3} min^{-1}, the stimulated steady state concentration of nuclear NF-κB approaches the unstimulated steady state concentration of about 0.46 nM.

The value of $k_{\beta\text{-TrCP}}$ also influences the stability of the stimulated steady states. A subsequent bifurcation analysis (see Section 2.7) reveals two Hopf bifurcation (HB) points at $k_{\beta\text{-TrCP}}$ values of approximately 0.033 min^{-1} and 0.094 min^{-1} (Figure 3.3A). For $k_{\beta\text{-TrCP}}$ values between these two bifurcation points, stable limit cycle oscillations of nuclear NF-κB can be observed. The limit cycle oscillations maintain a narrow range of period length from approximately 100 to 112 minutes for different values of $k_{\beta\text{-TrCP}}$ (Figure 3.3B). In contrast, the minimal and maximal values of the amplitude of nuclear NF-κB oscillations vary strongly (Figure 3.3A; dotted lines).

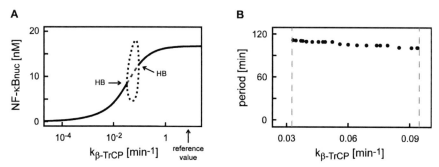

Figure 3.3: Bifurcation analysis of stimulated steady state of nuclear NF-κB with respect to $k_{\beta\text{-TrCP}}$.
(A) The value of $k_{\beta\text{-TrCP}}$ influences the stimulated steady state concentration of nuclear NF-κB and the stability of this steady state. The thick black lines denote stable stimulated steady states of nuclear NF-κB (NF-κBnuc). The grey dashed line denotes instable stimulated steady states. The dotted lines indicate minimal and maximal values of the amplitude of stable limit cycle oscillations occurring between the two Hopf bifurcation (HB) points at approximately 0.033 min⁻¹ and 0.094 min⁻¹. The reference parameter value of $k_{\beta\text{-TrCP}}$, which coincides with the published values of k_8 and k_{10}, is indicated at 6 min⁻¹. (B) Calculated period length of the limit cycle oscillations simulated for selected $k_{\beta\text{-TrCP}}$ values between the Hopf bifurcation points (indicated by dashed grey lines). The period hardly changes for the considered values of $k_{\beta\text{-TrCP}}$.

3.2.3 The value of $k_{\beta\text{-TrCP}}$ influences signal amplitude and signalling time of nuclear NF-κB dynamics in response to TNF stimulation

Next, the impact of $k_{\beta\text{-TrCP}}$ on the transient dynamics of nuclear NF-κB (NF-κBnuc) upon TNF stimulation is investigated. For the reference value of $k_{\beta\text{-TrCP}}$ of 6 min⁻¹, the transient dynamics of nuclear NF-κB upon TNF stimulation is shown in Figure 3.4A. In the first 90 min after TNF stimulation, a transient strong increase in the nuclear NF-κB concentration (first peak) with an amplitude of approximately 56.1 nM is observed. This first peak is followed by four minor peaks with decreasing amplitudes indicating damped oscillations of nuclear NF-κB concentration upon TNF stimulation (Figure 3.4A). In addition to the dynamics for the reference value of $k_{\beta\text{-TrCP}}$, nuclear NF-κB dynamics are shown for the reduced $k_{\beta\text{-TrCP}}$ value of 0.12 min⁻¹ (Figure 3.4B) and a strongly reduced value of 10^{-3} min⁻¹ (Figure 3.4C). Comparing the nuclear NF-κB dynamics in Figure 3.4A and B reveals that the oscillations loose damping if the value of $k_{\beta\text{-TrCP}}$ is reduced from 6 min⁻¹ to 0.12 min⁻¹. To analyse this change in the dynamics, the signalling time (Section 2.4) of the nuclear NF-κB dynamics is calculated for different values of $k_{\beta\text{-TrCP}}$ between 10^{-4} min⁻¹ and 10 min⁻¹ (Figure 3.4E). The simulations demonstrate that the closer $k_{\beta\text{-TrCP}}$ approaches the values of the Hopf bifurcation points (0.033 min⁻¹ and 0.094 min⁻¹, indicated by dashed grey lines in Figure 3.4E), the larger is the signalling time. For instance, considering the Hopf bifurcation point at 0.094 min⁻¹, the

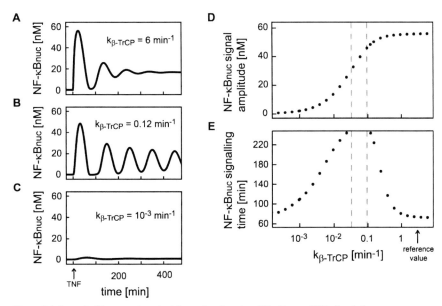

Figure 3.4: Impact of $k_{\beta\text{-TrCP}}$ on transient dynamics of nuclear NF-κB upon TNF stimulation.
Signal amplitude and signalling time are strongly affected by the value of $k_{\beta\text{-TrCP}}$. (A - C) Simulations of the dynamics of nuclear NF-κB (NF-κBnuc) upon TNF stimulation are shown for (A) the reference value of $k_{\beta\text{-TrCP}}$ (6 min⁻¹), (B) a reduced value of $k_{\beta\text{-TrCP}}$ (0.12 min⁻¹), and (C) a strongly reduced value of $k_{\beta\text{-TrCP}}$ (10^{-3} min⁻¹). (D) Dependence of the signal amplitude of nuclear NF-κB (NF-κBnuc) dynamics upon TNF stimulation for $k_{\beta\text{-TrCP}}$ values between 10^{-4} min⁻¹ and 10 min⁻¹. Dashed lines mark the $k_{\beta\text{-TrCP}}$ values of the two Hopf bifurcation points (see Figure 3.3A). (E) Impact of $k_{\beta\text{-TrCP}}$ on the signalling time of the nuclear NF-κB dynamics upon TNF stimulation. Dashed lines mark the $k_{\beta\text{-TrCP}}$ values of the two Hopf bifurcation points (see Figure 3.3A).

signalling time is increasing from about 72 min for $k_{\beta\text{-TrCP}} = 6$ min^{-1} to about 243 min for $k_{\beta\text{-TrCP}} = 0.12$ min^{-1}. Along with the signalling time also signal duration (Section 2.4) of the nuclear NF-κB dynamics increase from about 80 min for $k_{\beta\text{-TrCP}} = 6$ min^{-1} to about 190 min for $k_{\beta\text{-TrCP}} = 0.12$ min^{-1} (not shown).

Comparing the nuclear NF-κB dynamics in Figure 3.4A and C reveals that damped oscillations of nuclear NF-κB concentration occur upon TNF stimulation for the reference values of $k_{\beta\text{-TrCP}}$ (6 min^{-1}), while for the strongly reduced value (10^{-3} min^{-1}) the nuclear NF-κB concentration hardly reacts to TNF stimulation. To characterise this change in the transient dynamics of nuclear NF-κB, the signal amplitude is calculated for different values of $k_{\beta\text{-TrCP}}$ between 10^{-4} min^{-1} and 10 min^{-1} (Figure 3.4D). The simulations demonstrate that decreasing values of $k_{\beta\text{-TrCP}}$ reduce the signal amplitude of the nuclear NF-κB dynamics upon TNF stimulation from a maximal value of approximately 56.1 nM to a minimal value of about 0.46 nM (Figure 3.4D).

Taken together, signal amplitude, signalling time, and signal duration are strongly affected by value of $k_{\beta\text{-TrCP}}$. If $k_{\beta\text{-TrCP}}$ approaches the Hopf bifurcation points, the damping of the nuclear NF-κB dynamics is successively reduced, which is reflected by an increase of signalling time and signal duration (Figure 3.4E). The signal amplitude of the nuclear NF-κB dynamics upon TNF stimulation decreases by decreasing values of $k_{\beta\text{-TrCP}}$ (Figure 3.4D). For strongly reduced values of $k_{\beta\text{-TrCP}}$, the nuclear NF-κB concentration hardly changes upon TNF stimulation (Figure 3.4C).

3.3   Discussion

In this chapter, potential effects of the transcriptional regulation of β-TrCP abundance on nuclear NF-κB dynamics in response to TNF stimulation were investigated. In the modelling approach, changes of the rate constant $k_{\beta\text{-TrCP}}$ were assumed to represent variations in β-TrCP abundance (see Section 3.1). Under this assumption the model analysis predicted that changes in β-TrCP abundance influence the stimulated steady state of nuclear NF-κB (NF-κBnuc) (Figure 3.3A). This prediction is in agreement with many experimental observations. For instance, overexpression experiments showed that up-regulation of β-TrCP accelerates the proteasome-dependent degradation of IκB and thereby enhances the activity of NF-κB in various mammalian cell types (Bhatia et al., 2002; Fuchs et al., 1999; Kroll et al., 1999;

Muerkoster et al., 2005; Spiegelman et al., 2000; Spiegelman et al., 2001; Wang et al., 2004). In addition, expression of a dominant negative mutant of β-TrCP was reported to inhibit NF-κB activation and NF-κB-dependent transcription (Fuchs et al., 1999; Kroll et al., 1999; Soldatenkov et al., 1999; Spencer et al., 1999). This experimental observation is consistent with the finding that $k_{\beta\text{-TrCP}}$ values smaller than 10^{-3} min^{-1} reduce the stimulated steady state concentration to the unstimulated steady state concentration of NF-κB (Figure 3.3A). The simulations furthermore revealed that for $k_{\beta\text{-TrCP}}$ values greater than 1 min^{-1} the stimulated steady state of nuclear NF-κB remains almost constant (Figure 3.3A). This was expected since the two degradation rate constants k_8 and k_{10}, from which $k_{\beta\text{-TrCP}}$ was derived, have originally been set to sufficiently large values (6 min^{-1}) such that small deviations in these parameters hardly change the signalling dynamics of nuclear NF-κB (Lipniacki et al., 2004). This insensitivity of the stimulated steady state of nuclear NF-κB towards small deviations of $k_{\beta\text{-TrCP}}$ was also observed in the sensitivity analysis for the reference value of $k_{\beta\text{-TrCP}}$ (6 min^{-1}) (Section 3.2.1). The sensitivity analysis showed for the reference value of $k_{\beta\text{-TrCP}}$ that the binding of IKK to IκB/NF-κB (k_9) rather than the subsequent IκB degradation ($k_{\beta\text{-TrCP}}$) is the step that exerts most control on the IKK-dependent IκB degradation process (Figure 3.2). In the case of a reduced $k_{\beta\text{-TrCP}}$ value, however, the sensitivity of the stimulated steady state of nuclear NF-κB towards small changes of $k_{\beta\text{-TrCP}}$ was increased. These observations are of interest, since the most influential step in IKK-dependent IκB degradation is still unknown. Generally, phosphorylation of IκBα by IKKβ has been assumed to be the so called "rate limiting step" in the degradation process (Kanarek and Ben-Neriah, 2012). "Rate limiting step" in this context means that changes in the rate of this particular reaction most strongly influences the whole process of IκB degradation. The notion that IκB phosphorylation represents the "rate limiting step" was based on biochemical studies in a cell-free system, however, the "rate limiting step" might as well vary in different physiological contexts (Kanarek and Ben-Neriah, 2012). Alternative suggestions have placed the "rate limiting step" at the level of ubiquitination (Fuchs et al., 2004) or proteasome activation (Yang et al., 2001). Taking the β-TrCP over-expression experiments mentioned above into account (Bhatia et al., 2002; Fuchs et al., 1999; Kroll et al., 1999; Muerkoster et al., 2005; Spiegelman et al., 2000; Spiegelman et al., 2001; Wang et al., 2004), the presented findings of the steady state analyses collectively point towards a "rate limiting step" at IκB ubiquitination (Reaction 8 and Reaction 10 in Figure 3.1) in the case of small β-TrCP expression levels. This result is of special interest, since the most influential reactions may represent putative pharmacological intervention points.

In addition to the impact of β-TrCP abundance on the stimulated steady state of nuclear NF-κB, the analysis also predicts β-TrCP abundance to affect transient nuclear NF-κB dynamics in the immediate response to TNF stimulation. For instance, the signal amplitude of the nuclear NF-κB response directly correlated with the value of $k_{β-TrCP}$, i.e., the greater $k_{β-TrCP}$ the larger the signal amplitude (Figure 3.4D). This result contradicts the findings of earlier theoretical analyses (Cheong et al., 2006; Ihekwaba et al., 2004; Joo et al., 2007; Mathes et al., 2008) of the detailed kinetic model of canonical NF-κB signalling (Lipniacki et al., 2004) and related models (Hoffmann et al., 2002). In these publications, it has been predicted by means of sensitivity analyses that parameters of reactions governing the IKK-dependent degradation of IκB had hardly an impact on the first peak amplitude (Cheong et al., 2008; Mathes et al., 2008), which is identical to the signal amplitude considered in this chapter. This deviation between the published findings and those presented in Section 3.2.3, probably resulted from the specific choice of the parameter values in the IKK-dependent IκB degradation reactions (Reaction 8 and Reaction 10 in Figure 3.1). As seen in the sensitivity analysis in Section 3.2.1, the IKK-dependent IκB degradation rate constant $k_{β-TrCP}$ gains influence on the stimulated steady state concentration of nuclear NF-κB if $k_{β-TrCP}$ parameter values are chosen smaller than the originally published value of 6 min^{-1} (Figure 3.2).

The simulations in Figure 3.4D demonstrated that diminished concentrations of β-TrCP strongly reduce the signal amplitude of the nuclear NF-κB dynamics upon TNF stimulation. In the extreme case of strongly reduced β-TrCP abundance, nuclear NF-κB concentrations hardly changed upon TNF stimulation (Figure 3.4C). In consideration of these simulation results, it is interesting to note that Wnt/β-catenin signalling was reported to reduce the expression of HOS (Spiegelman et al., 2002b), which is a paralogue of β-TrCP commonly expressed in wild type cells (Saitoh and Katoh, 2001; Spiegelman et al., 2002b). Thus, it is tempting to speculate that Wnt/β-catenin signalling could decrease the signal amplitude of nuclear NF-κB and may thereby prevent canonical NF-κB signalling through transcriptional regulation of HOS. In support of this speculation, it has been reported that Wnt/β-catenin signalling influences canonical NF-κB activity via transcriptional regulation of β-TrCP abundance in human embryonic kidney 293T cells and mouse embryonic fibroblast NIH3T3 cells (Spiegelman et al., 2000; Spiegelman et al., 2002b).

There is an on-going debate in the field of NF-κB research whether oscillations of nuclear NF-κB occur *in vivo*, and if so, what functional and physiological importance they may have in the control of target gene expression (Barken et al., 2005; Nelson et al., 2005; Nelson et al., 2004; Tay et al., 2010). It has been speculated that number, period, and amplitude of oscillation peaks may determine the functional consequences of NF-κB signalling (Nelson et al., 2004). In this debate, the biological literature usually defines a consecutive sequence of three or more concentration peaks as oscillations. Depending on the number of observed peaks, so-called damped or sustained oscillations are discussed. The available data hardly allows drawing conclusions about the existence of limit cycle oscillations. In the presented theoretical analysis of the detailed kinetic model of canonical NF-κB signalling, the dynamics of nuclear NF-κB showed damped oscillations upon TNF stimulation for the reference value of $k_{\beta\text{-TrCP}}$ leading to a stable stimulated steady state of nuclear NF-κB. In addition, the bifurcation analysis of the stimulated steady states of nuclear NF-κB with respect to $k_{\beta\text{-TrCP}}$ revealed two Hopf bifurcation points (Figure 3.3A) at which the stable steady states of nuclear NF-κB concentration convert into stable limit cycle oscillations of nuclear NF-κB concentration for even small deviations in the value of $k_{\beta\text{-TrCP}}$. This finding implies that β-TrCP abundance may determine whether damped or sustained oscillations of the nuclear NF-κB concentration occur upon TNF stimulation in cells. The period length of the limit cycle oscillations remained almost constant at approximately 100 to 110 minutes for different values of $k_{\beta\text{-TrCP}}$ (Figure 3.3B). This period length is in agreement with experimentally measured inter-peak intervals in mouse fibroblasts (Hoffmann et al., 2002; Nelson et al., 2004; Tay et al., 2010). In contrast to the almost constant period length, the amplitude of the limit cycle oscillations strongly depended on the value of $k_{\beta\text{-TrCP}}$ (Figure 3.3A). This suggests that, β-TrCP abundance can also influence the amplitude of sustained oscillations of nuclear NF-κB in cells (Figure 3.3A), similarly to its impact on the signal amplitude of the damped oscillations (Figure 3.4A).

The presented results demonstrated that the regulation of β-TrCP abundance may affect steady states of nuclear NF-κB as well as the amplitude, signalling time, and signal duration of nuclear NF-κB dynamics upon TNF stimulation. These results are of importance, since β-TrCP is frequently up-regulated in human cancer and has been speculated to support tumourigenesis through activation of NF-κB-dependent anti-apoptotic pathways (Fuchs et al., 2004). Consequently, β-TrCP has been proposed to be a promising pharmacological target in cancer treatment (DiDonato et al., 2012; Fuchs et al., 2004; Kanarek and Ben-Neriah, 2012).

However, the sensitivity analysis (Section 3.2.1) predicted that interference with β-TrCP-dependent IκB ubiquitination (i.e. decreasing $k_{\beta\text{-TrCP}}$) may only be affective in cells having low β-TrCP concentration levels (Figure 3.2, grey bars). In cells with increased β-TrCP expression levels, the stimulated steady state of nuclear NF-κB was insensitive to changes in $k_{\beta\text{-TrCP}}$ (Figure 3.2, black bars). Consequently, pharmacological targeting of β-TrCP-dependent ubiquitination of IκB may not decrease the stimulated steady state of nuclear NF-κB in cancer cells with up-regulated β-TrCP expression. The analysis furthermore showed that interference with β-TrCP-dependent IκB ubiquitination (i.e. decreasing $k_{\beta\text{-TrCP}}$) may also reduce the regulatory influence of the A20 feedback but not the IκB feedback on the stimulated steady state of nuclear NF-κB (Figure 3.2). This is a counterintuitive result because β-TrCP regulates the ubiquitination of IκB and thereby IκB abundance. Hence, β-TrCP was expected to affect the regulatory activity of IκB in the IκB feedback rather than controlling the impact of A20. In more general terms, this result implies that local pharmacological interventions in one specific feedback may very well invoke changes in the regulatory impact of other distinct feedback mechanisms in a signalling pathway network. In this way, undesired and unforeseen side-effects can be readily provoked. Taken together, the results of the sensitivity analysis demand for caution, if β-TrCP-dependent IκB ubiquitination is intended to be pharmacologically targeted in cancer treatment.

In addition to cancer, aberrant activation of canonical NF-κB signalling is also detected in other severe human diseases, such as autoimmunity and inflammation (Ben-Neriah and Karin, 2011; Hayden and Ghosh, 2012; MacDonald et al., 2009). To reduce the stimulated steady state of nuclear NF-κB, the sensitivity analysis (Section 3.2.1) suggests two straight forward strategies. One approach would be to inhibit the formation of the IKK/IκB/NF-κB complex (k_9), which reduced the stimulated steady state of nuclear NF-κB independent of β-TrCP abundance albeit to different extent (Figure 3.2). The second approach is to generally inhibit protein and/or mRNA degradation. A decrease in the rate constants of the mRNA and protein degradation rates k_{19}, k_{23}, and k_{25} strongly reduced the stimulated steady state of nuclear NF-κB (Figure 3.2). A decrease in the remaining degradation rate constants k_2, k_5, k_6, k_{15}, k_{21}, and $k_{\beta\text{-TrCP}}$ (depending on its actual value) hardly affected it. These results indicate that the general inhibition of protein and/or mRNA degradation may reduce the stimulated steady state of nuclear NF-κB independent of β-TrCP abundance. Although different strengths of inhibition are needed to obtain a similarly efficient reduction for different β-TrCP concentrations. Indeed, inhibition of proteasomal IκB degradation has already been successfully used in the clinics to reduce NF-κB signalling in disease (Frankland-Searby and Bhaumik, 2012; Kanarek and Ben-Neriah, 2012). Administration of proteasome inhibitors, such as Bortezomib (reviewed in (Frankland-Searby and Bhaumik, 2012)), resulted in apoptosis of cancer cells. However, as the proteasome is directly or indirectly involved in almost all cellular processes, it is evident that its continuous inhibition is toxic.

All together, the results of this chapter show that the regulation of β-TrCP abundance may play an important role in the cell to control nuclear NF-κB dynamics. In fact, it has been demonstrated in human embryonic kidney 293T cells and mouse embryonic fibroblast NIH3T3 cells that Wnt/β-catenin signalling can affect canonical NF-κB activation through the regulation of β-TrCP abundance (Spiegelman et al., 2000; Spiegelman et al., 2002b). The results of this chapter furthermore indicate that β-TrCP may not represent a promising drug target to reduce NF-κB signalling, despite its advantage to be more substrate-specific compared to proteasome inhibitors (DiDonato et al., 2012; Frankland-Searby and Bhaumik, 2012).

4.   Dissection of the differential impact of FWD1 and HOS feedback on Wnt/β-catenin signalling

The two paralogues of β-TrCP, HOS and FWD1, have identical biochemical properties and are considered to function redundantly in the Wnt/β-catenin signalling pathway (see Section 1.4.2). Despite their similarities, they are differentially regulated by Wnt/β-catenin signalling. While HOS is inhibited by Wnt/β-catenin signalling, FWD1 is up-regulated (Section 1.4.2). This chapter focusses on the question whether the opposite control of FWD1 and HOS expression by Wnt/β-catenin signalling has distinct consequences on β-catenin/TCF dynamics or whether FWD1 and HOS fulfil redundant functions in the Wnt/β-catenin signalling pathway. To address this question, the detailed kinetic model of the Wnt/β-catenin signalling pathway (Lee et al., 2003) (see Section 2.2) is extended in Section 4.1 by the two transcriptional feedback mechanisms acting via HOS and FWD1. The modelling approach allows dissecting the specific effects of each individual feedback. A major result of this investigation is that Wnt/β-catenin signalling dynamics in wild type cells are primarily affected by the HOS feedback while the FWD1 feedback may establish a protection mechanism against loss of HOS expression. Thus, the two paralogues likely fulfil distinct functions in the regulation of Wnt/β-catenin signalling.

4.1   Derivation of a Wnt/β-catenin pathway model including the FWD1 and HOS feedbacks

Starting point for the derivation of a HOS and FWD1 feedback model is the detailed kinetic model of the Wnt/β-catenin signalling pathway introduced in Section 2.2. This detailed model is modified and extended by two transcriptional feedback mechanisms acting via HOS and FWD1. This extended model will be referred to as the two-feedback model in this thesis. In the two-feedback model, the HOS and FWD1 feedback mechanisms are implemented in such a way that the following three experimental observations are obeyed. First, FWD1 and HOS have identical biochemical properties in controlling β-catenin stability (Frescas and Pagano, 2008; Guardavaccaro et al., 2003; Nakayama et al., 2003). Second, while FWD1 is indirectly up-regulated by Wnt/β-catenin signalling via the induction of CRD-BP, the transcription of HOS is directly inhibited by the pathway (Noubissi et al., 2006; Spiegelman et al., 2000; Spiegelman et al., 2002b). Third, both paralogues are low expressed (Fuchs et al., 2004) and show an opposite expression pattern in normal and tumour tissues (Fuchs et al., 2004; Saitoh and Katoh, 2001; Spiegelman et al., 2002b) (see also Section 1.4.2).

4.1.1 Structure of the two-feedback model

To extend the detailed model by the feedback mechanisms three structural modifications need to be introduced. First, transcriptional regulation has to be considered. Second, the process of ubiquitination is included into the destruction core cycle. Third, the conservation relations of APC, GSK3, and TCF are revisited.

Incorporation of transcriptional regulation

The detailed model of Wnt/β-catenin signalling was based on experimental data of cellular extracts that lack transcriptional activity. To incorporate transcriptional regulation, a proposed modification has been to add up free and TCF-bound β-catenin in a pool of "transcriptionally available β-catenin" (Cho et al., 2006; Kim et al., 2007b; Wawra et al., 2007). This approach neglects the delay due to nuclear translocation of β-catenin before it associates with the TCF transcription factors. In addition, the fraction of TCF-bound β-catenin in the pool of "transcriptionally available β-catenin" is very small. In the two-feedback model a different strategy is followed. First, only β-catenin/TCF is considered as the transcriptionally active readout of the pathway. Second, the conservation law of the TCF concentration in the detailed model of Wnt/β-catenin signalling is replaced by a production and degradation mechanism of free TCF (Reaction 50 and Reaction 51, respectively, in Figure 4.1). In this approach, the concentration of transcriptionally active β-catenin/TCF complex is proportional to the free β-catenin concentration.

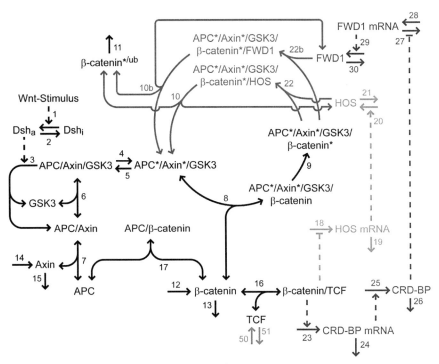

Figure 4.1: Reaction scheme of the two-feedback model.

The reaction scheme underlying the detailed model of Wnt/β-catenin signalling (black; Reactions 1 - 9 and Reactions 11 - 16) is extended by the HOS feedback (green; Reactions 18 - 21) and FWD1 feedback (blue; Reactions 23 - 30). HOS and FWD1 independently bind to the destruction complex to mediate the ubiquitination of phosphorylated β-catenin (red; Reaction 10, Reaction 10b, Reaction 22, and Reaction 22b). The newly introduced TCF production and degradation mechanisms are coloured in grey (Reaction 50 and Reaction 51, respectively). One-headed arrows denote reactions taking place in the indicated direction. Double-headed arrows illustrate binding reactions. Dashed arrows represent activation and dashed 'T's denote inhibition. Components in a complex are separated by a slash. The asterisk marks phosphorylation; ub stands for ubiquitinated species. The number next to an arrow specifies the number of the reaction. Reaction rates and parameters of the model are given in Appendix 7.3. A detailed description of the model is provided in the text of Section 4.1.

Association of HOS and FWD1 with the destruction complex

HOS and FWD1 both bind to the destruction complex (Kitagawa et al., 1999; Major et al., 2007) immediately after the phosphorylation of β-catenin by GSK3 (Liu et al., 1999; Su et al., 2008). The recognition of phosphorylated β-catenin by HOS or FWD1 is required for β-catenin ubiquitination, which targets β-catenin for proteasomal degradation (Spiegelman et al., 2000). Based on this experimental evidence, the destruction core cycle of the detailed model of Wnt/β-catenin signalling is extended by two independent binding reactions of HOS and FWD1 to the destruction complex (Reaction 22 and Reaction 22b, respectively, in Figure 4.1) after β-catenin phosphorylation by GSK (Reaction 9).

Due to the introduction of the novel APC*/Axin*/GSK3/β-catenin*/FWD1 and APC*/Axin*/GSK3/β-catenin*/HOS complexes in the destruction core cycle, the conservation relations of GSK3 and APC have to be adopted. In the two-feedback model, the conservation relation of GSK3 sums up free GSK3 and all GSK3-containing complexes to a constant total concentration of GSK3 (Equation 7.105, Appendix 7.3). Accordingly, the conservation relation for APC includes free APC and all APC-containing complexes (Equation 7.106, Appendix 7.3.2).

4.1.2 Rate kinetics of the two-feedback model

In general, constant production rates are assumed in the two-feedback model. In the case of all other processes, mass action kinetics is assumed (specified in Appendix 7.3). The only exceptions to that rule are the transcription of CRD-BP mRNA (Reaction 23 in Figure 4.1), the degradation of the FWD1 mRNA (Reaction 27), and the inhibition of HOS mRNA production (Reaction 18).

The transcription of CRD-BP mRNA is induced by cooperative binding of β-catenin/TCF complexes to two binding sites in the CRD-BP promoter (Gu et al., 2008; Noubissi et al., 2006). To describe this transcriptional regulation mechanism, a second order Hill-type function is used (Equation 7.131, Appendix 7.3). In addition, a constant basal transcription rate v_{basal} is considered.

The inhibition of the FWD1 mRNA degradation by CRD-BP and the inhibition of HOS mRNA expression by β-catenin/TCF complexes are both modelled by second order Hill-type inhibition functions (Equation 7.135 and Equation 7.125, respectively, in Appendix 7.3).

4.1.3 Definition of feedback strength

The transcription of HOS mRNA and the degradation of FWD1 mRNA (Figure 4.1; Reaction 18 and Reaction 27, respectively) are modelled by inhibition functions (Equation 7.125 and Equation 7.135, respectively, in Appendix 7.3). The inhibition constants

k_i and k_{i2} modulate the repressive impact of the β-catenin/TCF complex on HOS mRNA production and of CRD-BP protein on FWD1 mRNA degradation, respectively. HOS feedback strength is defined as the inverse of k_i and FWD1 feedback strength is defined as the inverse of k_{i2} (Equation 4.1 and Equation 4.2, respectively).

$$\text{HOS feedback strength } [\text{nM}^{-1}] = \frac{1}{k_i} \qquad\qquad 4.1$$

$$\text{FWD1 feedback strength } [\text{nM}^{-1}] = \frac{1}{k_{i2}} \qquad\qquad 4.2$$

For each feedback it holds that an increasing inhibition constant lowers the repressive impact and thus reduces the feedback strength. In the limit case that the inhibition constant approaches infinity, the feedback strength is zero. Consequently, changes in the concentration of the β-catenin/TCF complex (and CRD-BP) have no impact on the concentrations of HOS mRNA and FWD1 mRNA. In the model analysis, neither inhibition constant (k_i nor k_{i2}) can be set to infinity for numerical reasons. Rather, k_i and k_{i2} are set to the sufficiently large value of 10^6 nM, which corresponds to the feedback strength of 10^{-6} nM^{-1}. Retaining HOS or FWD1 feedback strength at this fixed value of 10^{-6} nM^{-1} renders HOS or FWD1 expression almost independent of changes in the concentration of the β-catenin/TCF complex. That is, the respective feedback regulation is disabled. It follows that the concentrations of HOS and FWD1 are retained fixed at their initial steady state values if the HOS and FWD1 feedback, respectively, are disabled in the simulations.

Taken together, it is important to remember that the term disabled HOS feedback (or disabled FWD1 feedback) means hereafter that the HOS (or FWD1) feedback strength is set to the value of 10^{-6} nM^{-1}, which fixes the concentrations of HOS (or FWD1) at its initial steady state value and abolishes the regulatory impact of the β-catenin/TCF complex on HOS (or FWD1) concentration.

To distinguish between the individual effects of the HOS and FWD1 feedback on Wnt/β-catenin signalling, the analysis in Section 4.2 and Section 4.3 starts by disabling the FWD1 feedback while varying the HOS feedback strength and then, *vice versa*, disabling the HOS feedback while varying the FWD1 feedback strength. Later in Section 4.3, the effects of combinations of various feedback strengths of both feedbacks are investigated.

4.1.4 Parameterisation

Parameterisation of the reactions adopted from the detailed model of Wnt/β-catenin signalling

Most parameters of the original detailed model of Wnt/β-catenin signalling (Table 7.3, Table 7.4, and Table 7.5 in Appendix 7.2) are adopted by the two-feedback model (Table 7.7 and Table 7.8 in Appendix 7.3). In case of Reaction 7, Reaction 8, Reaction 16, and Reaction 17, the detailed model of Wnt/β-catenin signalling assumed fast equilibria of complex formation (Appendix 7.2). This assumption may not hold anymore in the two-feedback model due to its modifications (see Section 4.1.1). The feature of fast complex formation is still captured in the two-feedback model by choosing sufficiently large association rate constants k_7, k_8, k_{16}, and k_{17}. The rate constants of the corresponding dissociation reactions (k_{7r}, k_{8r}, k_{16r}, and k_{17r}) in the two-feedback model are adjusted in such a way that the dissociation constants K_7, K_8, K_{16}, and K_{17}, respectively, of the detailed model of Wnt/β-catenin signalling are maintained in the two-feedback model.

Parameterisation of TCF production and degradation

The degradation rate constant k_{51} of free TCF is estimated from TCF time series experiments (Zhang et al., 2010). The TCF production rate v_{50} is adjusted such that the TCF steady state concentration in the absence of a Wnt stimulus corresponds to that of the detailed model of Wnt/β-catenin signalling (Table 7.7, Appendix 7.3).

Parameterisation of the reactions involved in the ubiquitination process

Since substrate binding is thought to stabilise HOS and FWD1 (Davis et al., 2002; Li et al., 2004) the binding Reaction 22 and Reaction 22b are considered to be faster than the degradation of HOS and FWD1. To represent the identical biochemical properties of HOS and FWD1 (Frescas and Pagano, 2008), the binding rate constants k_{22} and k_{22b} as well as the ubiquitination and dissociation rates k_{10} and k_{10b} are set equal (Table 7.7, Appendix 7.3).

Parameterisation of the reactions in the HOS feedback

The maximal HOS transcription rate $vmax_{18}$ (Table 7.7, Appendix 7.3) is adjusted such that the maximal possible HOS protein concentration is low compared to the concentrations of the other protein species in the two-feedback model. The degradation rate k_{19} of HOS mRNA had been estimated by (Spiegelman et al., 2002b). Following the argumentation of Lipniacki and colleagues concerning an initial estimation of the translation rate based on the produced amino acid sequence length (Lipniacki et al., 2004), the translation rate constant k_{20} of HOS (having a length of 542 amino acids) is set to 30 proteins per minute and mRNA. The degradation rate constant k_{21} of HOS protein is estimated from time series experiments (Li et al., 2004).

Parameterisation of the reactions in the FWD1 feedback

In contrast to HOS, which is expressed in wild type cells, CRD-BP as well as FWD1 are expressed in colon cancer cells (Dimitriadis et al., 2007; Fuchs et al., 2004; Saitoh and Katoh, 2001; Spiegelman et al., 2002b). Colon cancer cells often harbour mutations in APC (Giles et al., 2003). These mutations are assumed to constitutively activate Wnt/β-catenin signalling resulting in a pathological deregulation of target gene expression as observed in the case of CRD-BP and FWD1. In order to include the FWD1 feedback into the two-feedback model, the model needs to be adapted to capture the characteristics of colon cancer cells. To simulate colon cancer cells, parameter sets describing different APC mutations have been published for the detailed model of Wnt/β-catenin signalling (Cho et al., 2006). To simulate the cancer cell situation in the two-feedback model the parameter set of the APC mutant "m7" (Table 7.9, Appendix 7.3.5) is chosen as a representative parameter set for an APC mutation (see Section 2.3).

To estimate the parameter values in Reaction 23 and Reaction 24, published CRD-BP mRNA half-life as well as copy number data (Schwanhausser et al., 2011) is used and a 15-fold higher copy number in APC mutant compared to wild type cells is assumed as reported by (Dimitriadis et al., 2007). The translation rate k_{25} and the CRD-BP protein degradation rate k_{26} are taken from (Schwanhausser et al., 2011).

The maximal degradation rate constant of FWD1 mRNA k_{27} (Table 7.7, Appendix 7.3) is adjusted such that the effective degradation rate v_{27} in case of a moderate FWD1 feedback strength (1 nM^{-1}) is in the range of 10^{-3} - 10^{-2} $nM \cdot min^{-1}$. This parameter range is estimated from FWD1 mRNA time course data (Elcheva et al., 2009; Spiegelman et al., 2000; Spiegelman et al., 2001). Analogous to the case of HOS, the translation rate constant k_{29} of FWD1 (having a length of 605 amino acids) is set to 30 proteins per minute and mRNA. The protein degradation rate of FWD1 k_{30} is taken from (Davis et al., 2002).

4.1.5 Pathway stimulation by Wnt

Three different stimulus conditions have been considered in the detailed model of Wnt/β-catenin signalling (Lee et al., 2003): (i) the unstimulated pathway in the absence of Wnt, (ii) pathway stimulation by a constant Wnt stimulus of 1, and (iii) a transient Wnt stimulation (Appendix 7.2.5). The stimulus conditions (i) and (ii) have been introduced to characterise a reference steady state of the unstimulated Wnt/β-catenin signalling pathway and a standard stimulated steady state of Wnt/β-catenin signalling. The stimulus condition (iii) of a transient Wnt stimulation has been used to simulate the transient dynamics of β-catenin. It takes into account that Wnt stimulation *in vivo* is transient, likely due to receptor inactivation and other down-regulatory processes (Lee et al., 2003).

The three different stimulus conditions (i), (ii), and (iii) are adopted in the two-feedback model (Appendix 7.3.4). In the unstimulated pathway, the value of Wnt equals zero (Equation 7.141, Appendix 7.3.4). In this case, the abundance of each species in the model is given by its unstimulated steady state concentration. The stimulated steady state concentrations upon constant or transient Wnt stimulation (Equation 7.142 and Equation 7.143, respectively, in Appendix 7.3.4) are determined by the respective values of Wnt in the limit of time t approaching infinity. In this limit, the value of Wnt equals 1 in the case of constant Wnt stimulation. In contrast, in the case of transient Wnt stimulation, the value of Wnt equals zero, because Equation 7.143 (Appendix 7.3.4) simplifies to zero in the limit of time t approaching infinity. This means that the steady state upon transient Wnt stimulation is equal to the unstimulated steady state. In consequence, all analyses concerning the steady state upon transient Wnt stimulation can be performed without integrating the ODE system to infinity in simulations. Rather, the unstimulated steady state can be calculated by setting the time derivative of each species concentration in the two-feedback model to zero and solving the resulting algebraic equation system for the species concentrations.

4.2 Analysis of the impact of the modifications and extensions in the two-feedback model

4.2.1 Comparison of dynamical features of the two-feedback model and the original detailed model of Wnt/β-catenin signalling

For the detailed model of Wnt/β-catenin signalling, the dynamics of total β-catenin in response to transient Wnt stimulation have been published (Lee et al., 2003). Total β-catenin concentration is given by the sum of the concentrations of all β-catenin species and β-catenin-containing complexes in the model. The dynamics of total β-catenin of the detailed model of Wnt/β-catenin signalling and the two-feedback model upon transient Wnt stimulation are shown in Figure 4.2. In order to compare the two-feedback model with the detailed model of Wnt/β-catenin signalling, the two feedbacks are both disabled (see Section 4.1.3) rendering HOS and FWD1 mRNA independent of the β-catenin/TCF complex concentration. The comparison of the simulations shows that the two-feedback model reproduces the dynamics of total β-catenin of the detailed model of Wnt/β-catenin signalling very well (Figure 4.2).

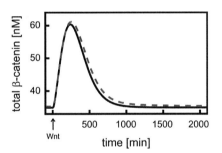

Figure 4.2: Impact of the modifications and extensions in the two-feedback model.
Comparison of the dynamics of total β-catenin in response to transient Wnt stimulation in the detailed model of Wnt/β-catenin signalling (Lee et al., 2003) (black line) and the two-feedback model (red dashed line). To allow for a comparison of the two models, the HOS and the FWD1 feedback are disabled (see Section 4.1.3) in the simulation. The two-feedback model reproduces the dynamics of total β-catenin of the detailed model of Wnt/β-catenin signalling very well.

The steady state concentrations of total β-catenin in the absence of Wnt are very similar in both models and deviate less than 1.2% (Table 4.1). In contrast, the steady state concentrations of total β-catenin upon constant Wnt stimulation differ between the models by about 1.2-fold (Table 4.1). This difference is due to the production and degradation mechanisms of TCF that are introduced in the two-feedback model.

Table 4.1: Comparison of steady state concentrations.
Comparison of steady state concentrations of total β-catenin and β-catenin/TCF complex between the two-feedback and the detailed model of Wnt/β-catenin signalling for different conditions of Wnt stimulation. The HOS and the FWD1 feedback of the two-feedback model are disabled (Section 4.1.3) in the simulations.

Wnt stimulation	total β-catenin		β-catenin/TCF	
	detailed model	two-feedback model	detailed model	two-feedback model
Unstimulated	34.9 nM	35.3 nM	6.8 nM	6.9 nM
Constant	177.8 nM	208.8 nM	12.5 nM	42.1 nM
Transient	34.9 nM	35.3 nM	6.8 nM	6.9 nM

In the detailed model of Wnt/β-catenin signalling, total β-catenin represented the readout of pathway activation, because it was the measureable quantity in the performed experiments. In the following sections, however, β-catenin/TCF complex and not total β-catenin is considered to be the biologically relevant readout of pathway activation, since this complex represents the transcriptionally active β-catenin species in the two-feedback model. Table 4.1 shows that the unstimulated steady state concentrations of the β-catenin/TCF complexes in the detailed model of Wnt/β-catenin signalling and the two-feedback model are almost equal. However, the two-feedback model produces an approximately 3.4-fold higher steady state concentration of the β-catenin/TCF complex upon constant Wnt stimulation than the detailed model of Wnt/β-catenin signalling. The underlying reason is again the introduction of a TCF production and degradation mechanism in the two-feedback model.

4.2.2 Comparison of the unstimulated steady state concentrations of FWD1 and HOS under wild type and APC mutant conditions

To check whether the two-feedback model can reproduce the reported expression pattern of the two β-TrCP paralogues (Spiegelman et al., 2002b), the unstimulated steady state concentrations of FWD1 and HOS are calculated in the two-feedback model simulating a wild type or APC mutant "m7" scenario (see Section 2.3). The qualitative comparison of unstimulated steady state concentrations of FWD1 and HOS in Figure 4.3 shows a very low expression of FWD1 and higher expression of HOS in the wild type ($7.3 \cdot 10^{-4}$ nM and 0.69 nM, respectively) but high expression of FWD1 and a very low expression of HOS in the APC mutant case (0.76 nM and $4.9 \cdot 10^{-4}$ nM, respectively). The simulations in Figure 4.3 consider HOS and FWD1 feedback strengths of 0.13 nM^{-1}. Other feedback strengths ranging from 0.06 nM^{-1} to 0.3 nM^{-1} were also simulated and yielded similar results. Taken together, the two-feedback model reproduces the reported opposite expression pattern of the two proteins (Spiegelman et al., 2002b).

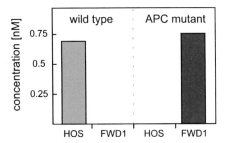

Figure 4.3: The two-feedback model reproduces the differential expression pattern of HOS and FWD1.
Unstimulated steady state concentrations of FWD1 and HOS are calculated in the two-feedback model in wild type and APC mutant "m7" scenario (see Section 2.3). They reproduce the differential expression patterns observed in experiments (Spiegelman et al., 2002b). HOS and FWD1 feedback strengths are set to 0.13 nM^{-1} in the simulations.

4.2.3 Variation of feedback strength allows for feedback-specific modulation of HOS or FWD1 expression

Next, the influence of feedback strength on the unstimulated steady state expression of HOS and FWD1 is analysed in wild type cells. Two cases are considered. First, the HOS feedback strength is varied while the FWD1 feedback is disabled (referred to as scenario I). Then, the HOS feedback is disabled and the FWD1 feedback strength is changed (i.e. scenario II).

The simulations in scenario I show that increasing HOS feedback strength decreases the unstimulated steady state concentration of HOS (Figure 4.4A). This is due to the repressive impact of β-catenin/TCF on HOS mRNA production, which increases with increasing HOS feedback strength. In contrast, variations of HOS feedback strength do not result in observable effects on the unstimulated steady state levels of FWD1 (Figure 4.4C).

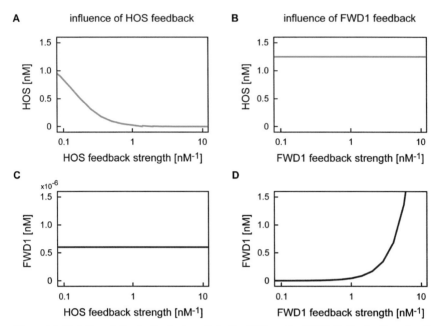

Figure 4.4: Variation of the feedback strength in the HOS or the FWD1 feedback allows for the independent modulation of HOS or FWD1 expression, respectively.

(A, C) HOS feedback strength is gradually increased, while the FWD1 feedback is disabled (scenario I). The unstimulated steady state concentration of HOS decreases in response to increasing HOS feedback strength (A), while FWD1 expression is not affected (C). (B, D) FWD1 feedback strength is gradually increased, while the HOS feedback is disabled (scenario II). The unstimulated steady state concentration of FWD1 rises in response to increasing FWD1 feedback strength (D), while the unstimulated steady state concentration of HOS stays constant (B).

In scenario II, increasing FWD1 feedback strength increases the unstimulated steady state levels of FWD1 (Figure 4.4D) due to the positive regulation of FWD1 by the β-catenin/TCF complex. The simulations furthermore show that a variation of the FWD1 feedback strength does not have an observable effect on the unstimulated steady state concentration of HOS (Figure 4.4B).

Taken together, the analysis demonstrates that variation of one feedback strength allows for feedback-specific modulation of either HOS or FWD1 expression. Note that very different unstimulated steady state concentrations of HOS and FWD1 are present (1.25 nM and $0.56 \cdot 10^{-6}$ nM, respectively) under wild type conditions if both feedbacks are simultaneously disabled. At last, it is again emphasised that the results relating to the analysis of the unstimulated steady states can also be obtained in the analysis of the stimulated steady states upon transient Wnt stimulation since unstimulated and transiently stimulated steady states are equal (see Section 4.1.5). Note that, in contrast to transient Wnt stimulation, the stimulated steady state upon constant Wnt stimulation differs from the unstimulated steady state.

4.3 Analysis of the two-feedback model

4.3.1 HOS feedback strength modulates the steady state concentration of the β-catenin/TCF complex

To investigate the impact of the HOS feedback on the steady state concentration of the β-catenin/TCF complex upon transient Wnt stimulation, scenario I (page 75) is revisited: the FWD1 feedback is disabled while the HOS feedback strength is varied. The simulations show that weak HOS feedbacks of less than 1 nM^{-1} maintain a low steady state concentration of the β-catenin/TCF complex upon transient Wnt stimulation, whereas strong HOS feedbacks greater than 10 nM^{-1} induce an accumulation of β-catenin/TCF complexes (Figure 4.5A). As shown in Figure 4.4A, strong HOS feedbacks reduce the HOS protein concentration. Consequently, the HOS-dependent β-catenin degradation is strongly inhibited resulting in the accumulation of β-catenin/TCF complexes. In the range of intermediate HOS feedback strength (1 - 10 nM^{-1}), multiple β-catenin/TCF steady state solutions are calculated. An analysis of the stability of the steady state solutions (Appendix 8.1) reveals that only the highest β-catenin/TCF concentration corresponds to a stable solution. Thus, steady state concentrations of the β-catenin/TCF complex react upon transient Wnt stimulation in a switch-like manner without hysteresis to changes in HOS feedback strength. Note again that an analysis of the unstimulated steady state of the β-catenin/TCF complex yields the same results. This implies in particular, that the steady state concentration of the β-catenin/TCF complex rises even in the absence of any Wnt stimulation if the HOS feedback strength increases above 1 nM^{-1}.

Next, the unstimulated and constantly stimulated steady state concentrations of the β-catenin/TCF complex are compared for two different HOS feedback strengths (Figure 4.5B). In case of a weak HOS feedback strength (0.3 nM^{-1}), the steady state concentration of the β-catenin/TCF complex upon constant Wnt stimulation is approximately 6-fold larger than that in the absence of Wnt (approximately 6.9 nM). In contrast, a high HOS feedback strength (30 nM^{-1}) renders the β-catenin/TCF steady state concentration unresponsive to Wnt stimulation. In this case, the β-catenin/TCF steady state obtains for all three stimulation conditions (unstimulated, transiently stimulated, and constantly stimulated) the highest concentration that is possible in the two-feedback model (approximately 442 nM).

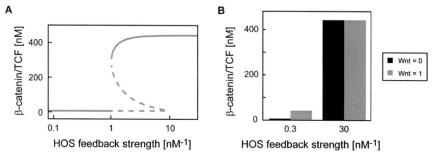

Figure 4.5: Steady state of the β-catenin/TCF complex depends on HOS feedback strength.
(A) The steady state concentration of the β-catenin/TCF complex responds upon transient Wnt stimulation in a step-like manner to increasing HOS feedback strength. The green lines denote stable steady state solutions; the grey dashed line marks unstable solutions (see Appendix 8.1 for a more detailed bifurcation analysis). (B) Unstimulated and constantly stimulated steady state concentrations of the β-catenin/TCF complex (black and grey bars, respectively) for a representative low and high HOS feedback strength (0.3 nM^{-1} and 30 nM^{-1}, respectively). The high HOS feedback strength of 30 nM^{-1} strongly increases the β-catenin/TCF steady state concentration and renders it unresponsive to Wnt stimulation. The FWD1 feedback is disabled in all simulations shown in the figure.

4.3.2 HOS feedback strength influences the dynamics of the β-catenin/TCF complex in response to transient Wnt stimulation

To investigate the impact of the HOS feedback strength on the dynamics of the β-catenin/TCF complex in response to transient Wnt stimulation, the FWD1 feedback is kept disabled (scenario I; see page 75). The simulations show that the HOS feedback strength strongly influences the dynamics of the β-catenin/TCF complex (Figure 4.6A). The impact on the dynamics is quantified by three measures that characterise typical properties of transient signal response dynamics: signalling time, signal duration as well as signal amplitude (see Section 2.4). These three measures are calculated for the transient increase in the concentration of the β-catenin/TCF complex considering different HOS feedback strengths. The simulations are restricted to low HOS feedback strengths since strong feedbacks render the β-catenin/TCF complex unresponsive to Wnt stimulation (Figure 4.5B). The analysis shows that increasing HOS feedback strength increases β-catenin/TCF signalling time, extends β-catenin/TCF signal duration, and enhances β-catenin/TCF signal amplitude (Figure 4.6B).

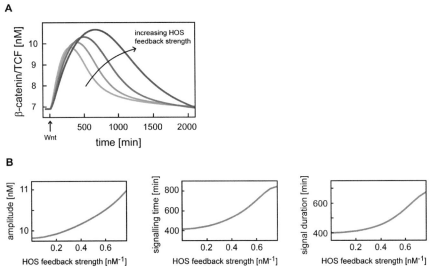

Figure 4.6: HOS feedback strength influences the dynamics of the β-catenin/TCF complex upon transient Wnt stimulation.

Increasing HOS feedback strengths increase signal amplitude, increase signalling time, and extend signal duration of the β-catenin/TCF dynamics upon transient Wnt stimulation. (A) The dynamics of the β-catenin/TCF complex upon transient Wnt stimulation for different HOS feedback strengths. Colour-code: disabled HOS feedback (light green), HOS feedback strength of 0.3 nM^{-1} (green), 0.5 nM^{-1} (darker green), and 0.7 nM^{-1} (dark green). The FWD1 feedback is disabled in all four cases. (B) Quantification of signal amplitude, signalling time, and signal duration of the dynamics of β-catenin/TCF complex (see Section 2.4 for definitions).

4.3.3 FWD1 feedback strength does not affect the dynamics of the β-catenin/TCF complex for the reported HOS expression level

To determine the regulatory influence of the FWD1 feedback on the β-catenin/TCF dynamics, scenario II (page 75) is now revisited: i.e., the HOS feedback is disabled while the FWD1 feedback strength is varied. Although variation of FWD1 feedback strength changes FWD1 concentration over a broad range (Figure 4.4D), no influence of FWD1 feedback strength on the steady state concentration of the β-catenin/TCF complex upon transient Wnt stimulation is observed (Figure 4.7A). Furthermore, β-catenin/TCF dynamics in response to transient Wnt stimulation are hardly affected (Figure 4.7B). Signalling time, signal duration, and signal amplitude of the β-catenin/TCF complex dynamics stay almost constant (Figure 4.7C). This insensitivity of the β-catenin/TCF complex dynamics is due to the present expression level of HOS (Figure 4.4B). This present HOS expression level is not changed by the β-catenin/TCF complex dynamics because the HOS feedback is disabled. The present HOS concentration compensates for reduced FWD1 concentrations in case of weak FWD1 feedbacks of less than 1 nM^{-1} (Figure 4.4D). It is interesting to note that in the case of strong FWD1 feedbacks (larger than 10 nM), the increased expression of FWD1 in addition to the present HOS concentration can also not accelerate the β-catenin/TCF dynamics (Figure 4.7B and middle panel of Figure 4.7C).

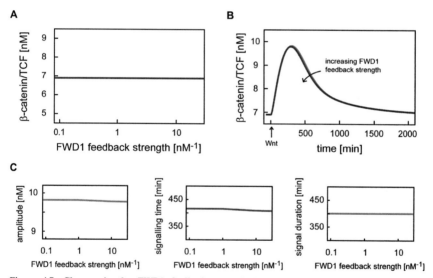

Figure 4.7: Changes in the FWD1 feedback strength hardly affect β-catenin/TCF steady state concentration and dynamics in combination with a disabled HOS feedback.

The value of the FWD1 feedback strengths hardly affects signal amplitude, signalling time, and signal duration of the β-catenin/TCF dynamics upon transient Wnt stimulation. (A) The steady state concentration of β-catenin/TCF complex upon transient Wnt stimulation for different FWD1 feedback strengths in combination with a disabled HOS feedback. (B) Overlay of β-catenin/TCF complex dynamics upon transient Wnt stimulation comparing the disabled FWD1 feedback (light blue) and a strong FWD1 feedback with feedback strength of 10 nM^{-1} (dark blue). The HOS feedback is disabled in both cases. (C) Quantification of signalling time, signal duration, and signal amplitude of the dynamics of the β-catenin/TCF complex (see Section 2.4 for definitions).

4.3.4 FWD1 feedback strength influences the steady state concentration of the β-catenin/TCF complex in the absence of HOS

As shown in Section 4.3.3, HOS expression can mask possible effects of the FWD1 feedback. Therefore, a new scenario of complete absence of HOS is investigated in this section. To model loss of HOS, the HOS mRNA production rate $vmax_{18}$ is set to zero. Loss of HOS strongly reduces β-catenin degradation resulting in an elevation of the steady state concentration of β-catenin/TCF complex from 6.8 nM to 442 nM even in the absence of Wnt stimulation (Figure 4.8A). Neither transient nor constant Wnt stimulation change this high steady state concentration of 442 nM. Increasing FWD1 feedback strength increases the concentration of FWD1 (Figure 4.4D), allowing for β-catenin degradation via the β-catenin*/APC*/Axin*/GSK3/FWD1 complex. Accordingly, the unstimulated steady state concentration of the β-catenin/TCF complex stays low for sufficiently strong FWD1 feedbacks (greater than 10^{-2} nM^{-1}) despite the absence of HOS (Figure 4.8B). For a FWD1 feedback strength of 0.13 nM for instance, the unstimulated steady state concentration of the β-catenin/TCF complex is approximately 7.5 nM, which deviates only 0.6 nM from the original unstimulated steady state of the two-feedback model given in Table 4.1. A second consequence of this FWD1-mediated β-catenin degradation is that the Wnt/β-catenin

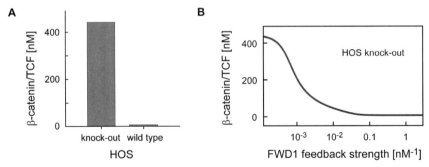

Figure 4.8: FWD1 feedback influences the steady state concentration of the β-catenin/TCF complex in the absence of HOS.

(A) In the absence of HOS, the steady state concentration of β-catenin/TCF complex increases to the maximal possible value in the two-feedback model. (B) The FWD1 feedback strength influences the unstimulated/transiently Wnt stimulated steady state concentration of the β-catenin/TCF complex once HOS is absent.

signalling pathway becomes responsive to transient as well as constant Wnt stimulation. In particular, in the scenario of combining loss of HOS with a FWD1 feedback of 0.13 nM feedback strength, the constantly stimulated steady state concentration of the β-catenin/TCF complex is about 42.4 nM. This is an about 5.7-fold increase compared to the unstimulated steady state demonstrating the responsiveness of the Wnt/β-catenin signalling pathway to stimulation.

4.3.5 Changes in the FWD1 or HOS expression level modulate the influence of the HOS feedback on the β-catenin/TCF dynamics

In the previous sections, the expression level of FWD1 was increased by enhancing FWD1 feedback strength (Figure 4.4D). Another mechanism to enhance FWD1 expression is to amplify FWD1 transcription. Experiments showed that the stress-induced JNK and the Akt/PKB signalling pathway regulate FWD1 expression (Spiegelman et al., 2000; Spiegelman et al., 2001). In the two-feedback model, transcription of FWD1 can be enhanced by increasing the FWD1 mRNA production rate v_{28}. The simulations show that higher FWD1 mRNA production rate accelerates the dynamics of the β-catenin/TCF complex upon transient Wnt stimulation (Figure 4.9A, blue line) compared to the dynamics of the original FWD1 mRNA production rate (black line). Similar to this decrease in the β-catenin/TCF signalling time, also the β-catenin/TCF signal duration is shortened and the signal amplitude is decreased for larger FWD1 mRNA production rates (Figure 4.9B, blue lines).

Similar to FWD1, transcriptional activation of HOS also affects the dynamics of the β-catenin/TCF complex to transient Wnt stimulation (Figure 4.9A, green line). The simulations show that increasing HOS mRNA production rate $vmax_{18}$ results in accelerated β-catenin/TCF complex dynamics, shortened signal duration, and decreased signal amplitude (Figure 4.9B, green lines). In contrast to FWD1, already a small increase of HOS transcription by less than 5-fold results in strong decreases in β-catenin/TCF signalling time, duration, and amplitude (Figure 4.9B).

Taken together, transcriptional up-regulation of FWD1 or HOS counteracts the impact of the HOS feedback strength on the dynamics of the β-catenin/TCF complex upon transient Wnt stimulation shown in Figure 4.6. In the extreme case of strong transcriptional up-regulation of FWD1 or HOS (greater than 30-fold), the impact of the HOS feedback can even be completely annihilated in the model simulations.

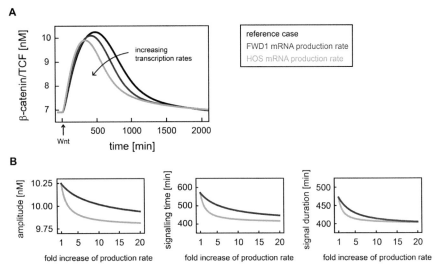

Figure 4.9: Transcriptional up-regulation of the concentration of FWD1 or HOS reduces the impact of the HOS feedback.

(A) Different response dynamics of the β-catenin/TCF complex to transient Wnt stimulation for selected combinations of FWD1 and HOS mRNA production rates. Black: HOS and FWD1 production rates of 10^{-6} nM·min^{-1} (reference case); blue: HOS and FWD1 production rates of 10^{-6} nM·min^{-1} and $5 \cdot 10^{-6}$ nM·min^{-1}, respectively; green: HOS and FWD1 production rates of $5 \cdot 10^{-6}$ nM·min^{-1} and 10^{-6} nM·min^{-1}, respectively. The HOS and FWD1 feedback strengths are set to 0.5 nM^{-1}. (B) Quantification of signalling time, signalling duration, and signal amplitude of the dynamics of the β-catenin/TCF complex upon transient Wnt stimulation for different values of FWD1 mRNA production rate (blue lines) or HOS mRNA production rate (green lines).

4.3.6 The FWD1 feedback mechanism does not protect against mutations of APC

The impact of the FWD1 and HOS feedback has been discussed so far in the case of wild type cells that are characterised by a much lower concentration of FWD1 than HOS. The analyses have shown that the impact of the FWD1 feedback was masked by the present HOS concentration under wild type conditions. In contrast, if HOS expression was lost (Section 4.3.4), the FWD1 feedback strength influenced the steady state concentration of the β-catenin/TCF complex. In contrast to wild type cells and similar to the loss of HOS scenario, primary tumours and cancer cell lines express hardly HOS but rather FWD1 (Fuchs et al., 2004; Saitoh and Katoh, 2001). Thus, FWD1 feedback effects might become observable under that condition, too.

An important human cancer type is colon cancer, which is the third most commonly diagnosed cancer in the world. In more than 85% of colon cancer samples, mutations in APC have been detected, which are paralleled by an aberrant activation of Wnt/β-catenin signalling (Giles et al., 2003). Parameter sets of APC mutants have been published for the detailed model of Wnt/β-catenin signalling (Cho et al., 2006). These parameter sets were adopted to simulate APC mutations within the two-feedback model (see Section 2.3). The simulations show that the unstimulated steady state concentrations of the β-catenin/TCF complex are increased in all APC mutants compared to the wild type scenario if the FWD1 and the HOS feedback are disabled (Figure 4.10B, black bars). Since overexpression of FWD1 was reported to promote the down-regulation of β-catenin in various cancer cell lines (Hart et al., 1999; Su et al., 2008), it is speculated that the FWD1 feedback may provide a protection mechanism against aberrant activation of Wnt/β-catenin signalling in cancer cells similar to its protective role in the case of loss of HOS expression (see Section 4.3.4). To explore this hypothesis, the unstimulated steady state concentrations of FWD1 under wild type and APC mutant conditions are calculated using the two-feedback model. In the simulations, the FWD1 and the HOS feedback are both set to low feedback strengths of 0.13 nM^{-1} (other feedback strengths lead to similar results). The fold increase of the unstimulated steady state concentration of FWD1 for each APC mutant with respect to the unstimulated steady state concentration of FWD1 in wild type cells is shown in Figure 4.10A. In all APC mutants, the steady state concentration of FWD1 is increased compared to wild type cells, despite the absence of a Wnt stimulus. Dependent on the APC mutant, the fold increase spans from approximately 2.6-fold for APC mutant "m1" to approximately 1100-fold for APC mutants "m9" to "m13". Next, the corresponding unstimulated steady state concentration of the β-catenin/TCF complex is calculated for the wild type and each APC mutant. The simulations show that the unstimulated steady state concentrations of the β-catenin/TCF complex hardly change if the FWD1 and HOS feedback strengths are set to 0.13 nM^{-1} (Figure 4.10B, grey bars) compared to those if both feedbacks are disabled (Figure 4.10B, black bars). The concentrations are similar despite the strong fold change of FWD1 expression under these conditions (Figure 4.10A). This result indicates that the FWD1 feedback mechanism probably does not protect Wnt/β-catenin signalling against mutations of APC.

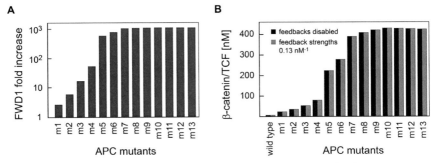

Figure 4.10: The FWD1 feedback mechanism does not protect against mutations of APC.
(A) Fold increase of FWD1 steady state concentrations of APC mutants "m1" to "m13" in the absence of Wnt stimulation with respect to the unstimulated steady state concentration of FWD1 in wild type cells. FWD1 and HOS feedback are both set to a representative feedback strength of 0.13 nM^{-1}. (B) The unstimulated steady state concentration of the β-catenin/TCF complex of wild type cells and APC mutants "m1" to "m13" in the case of disabled FWD1 and HOS feedbacks (black bars) compared to that in the case of FWD1 and HOS feedback strengths of 0.13 nM^{-1} (grey bars). For details on the simulation of APC mutants "m1" to "m13" see Section 2.3.

4.4   Discussion

In this chapter, the regulatory impact of the HOS and FWD1 feedback mechanism on the response dynamics of Wnt/β-catenin signalling was investigated. The modelling approach allowed distinguishing the influence of each individual feedback by modulation of their respective feedback strength. The analysis demonstrated that the opposite regulation of FWD1 and HOS expression levels by Wnt/β-catenin signalling results in distinct rather than redundant regulatory effects of the two feedbacks on the concentration of the β-catenin/TCF complex. It furthermore showed that the actual impact of each feedback strongly depends on the expression levels of FWD1 and HOS, in absolute concentrations and relative to each other.

In wild type cells, higher concentrations of HOS than FWD1 have been detected (Fuchs et al., 2004; Saitoh and Katoh, 2001; Spiegelman et al., 2002b). On the basis of this experimental finding, the model predicts that under wild type conditions primarily the HOS feedback influences the dynamics of the β-catenin/TCF complex in response to transient Wnt stimulation. The analysis showed that increasing HOS feedback strengths increased the steady state concentration of the β-catenin/TCF complex (Figure 4.5A). In the case of strong HOS feedbacks, the steady state concentration of the β-catenin/TCF complex reached the maximal possible concentration in the model (Figure 4.5B). Furthermore, the difference between the steady state concentration in the absence and in the presence of Wnt stimulation ceased (Figure 4.5B). This may indicate a hyper-activation of the β-catenin/TCF complex resulting in the pathologic activation of the signalling pathway independent of the presence or absence of a Wnt stimulus. Increasing HOS feedback strengths also increased the signalling time, signal duration, and amplitude of the dynamics of the β-catenin/TCF complex in response to transient Wnt stimulation (Figure 4.6). A HOS-feedback-strength-dependent stabilisation of Wnt/β-catenin pathway activation (i.e. increasing signal duration) has been experimentally shown in cells that express HOS but not FWD1 (Spiegelman et al., 2002b).

The model furthermore predicted that the FWD1 feedback by its self would not influence the β-catenin/TCF dynamics in wild type cells due to the reported HOS expression levels (Figure 4.7). This result is in line with the reported experimental observation that already minute levels of HOS compensate even extreme variations of FWD1 concentration (Nakayama et al., 2003). The analysis showed that the FWD1 feedback gained regulatory impact under the condition of increasing FWD1 concentration in combination with reduced HOS concentration. The extreme realisation of this condition is the case of loss of HOS expression. In this scenario, the degradation of β-catenin via the HOS-dependent mechanism is completely blocked. As a consequence, β-catenin/TCF concentrations strongly increased in the simulations even in the absence of Wnt stimulation (Figure 4.8A). The FWD1 feedback allowed for β-catenin degradation via the β-catenin*/APC*/Axin*/GSK3/FWD1 complex reducing the concentration of the β-catenin/TCF complex and keeping the pathway responsive to Wnt stimulation. In this way, the FWD1 feedback provided a protection mechanism against loss of HOS expression. This protection mechanism depended on the positive regulation of FWD1 expression by the Wnt/β-catenin signalling pathway.

It was speculated that the opposing transcriptional regulation of HOS and FWD1 may offset their individual impact on Wnt/β-catenin signalling, unless both paralogues would have different substrate specificity (Kanarek and Ben-Neriah, 2012). However, a bifurcation analysis (Appendix 7.4.5) demonstrated that, even for equal binding constants of both paralogues for phosphorylated β-catenin in the destruction complex, distinct dynamical effects may be expected. The specific effect, such as the number of steady states, depended on the actual combination of the values of HOS and FWD1 feedback strength (Figure 8.1B, Appendix 7.4.5). The analysis showed that at certain combinations of feedback strengths, stable limit cycle oscillations can occur in the model (Figure 8.2B, Appendix 7.4.5). Oscillations of Wnt/β-catenin signalling have been described in embryonic development (Aulehla and Herrmann, 2004; Aulehla and Pourquié, 2008; Dequeant et al., 2006; Goldbeter and Pourquie, 2008). Simulations showed that increasing FWD1 feedback strength reduced the period of the limit cycle oscillations (Figure 8.2B, Appendix 7.4.5). This result is of special interest because CRD-BP, which mediates the regulatory influence of the β-catenin/TCF complex on FWD1, is usually expressed in foetal tissues (Ioannidis et al., 2001). This may hint to a possible regulation of the FWD1 feedback strength in that context. One may thus speculate that the FWD1 feedback mechanism could be of relevance to tune the oscillation period of Wnt/β-catenin signalling during embryonic development.

So far, effects of increased expression levels of FWD1 due to changes in FWD1 feedback strength were discussed. Alternatively, expression levels of FWD1 can be elevated by transcriptional activation through other signalling pathways. Possible candidate pathways are the stress-induced JNK and Akt/PKB signalling pathways, which were reported to regulate FWD1 expression (Spiegelman et al., 2000; Spiegelman et al., 2001). The analysis showed that the regulation of FWD1 concentration by transcriptional activation could modulate the β-catenin/TCF dynamics and thus may influence Wnt/β-catenin target gene expression. One may speculate that another function of FWD1 may therefore be to allow for the modulation of the duration of β-catenin/TCF complex dynamics by other signalling pathways. In that case, the preferential expression of FWD1 or HOS in different cell types might possibly obey a specific functional demand of the cell, rather than simply result from their different regulation mechanisms. In one situation, a cell may need regulation of its Wnt response via the HOS feedback, while in another situation an integrative response to different environmental signals is preferred, established by transcriptional regulation of FWD1 expression by these signals.

The simulations showed that increased expression of FWD1 or HOS could counteract the regulatory impact of the HOS feedback on Wnt/β-catenin signal transduction (Figure 4.9). This has implications for experiments that rely on the expression of tagged-proteins to investigate the regulation of Wnt/β-catenin signalling. The artificial increase of FWD1 and/or HOS could strongly change the regulatory impact of the HOS feedback on β-catenin/TCF dynamics.

In colon cancer, mutations in APC are frequently detected in parallel to increased concentrations of β-catenin. In addition, FWD1 is often strongly expressed while HOS is down-regulated. Since increased FWD1 levels could reduce up-regulated β-catenin/TCF concentrations in case of loss of HOS expression in the simulations of wild type cells (Figure 4.8), it was investigated whether the FWD1 feedback may also provide a protection mechanism against up-regulation of β-catenin/TCF complex in cancer cells. The simulations showed that APC mutations led to elevated FWD1 expression levels due to the transcriptional control of FWD1 by the β-catenin/TCF complex (Figure 4.10A). However, the FWD1 feedback mechanism could not reduce the β-catenin/TCF steady state concentration in the mutant scenario to wild type concentrations again (Figure 4.10B). The reason is that mutations of destruction complex components already interfere with the phosphorylation of β-catenin, which is the prerequisite for HOS and FWD1 to act. Consequently, FWD1 or HOS seem to be poor targets to regulate β-catenin expression levels in these mutations.

5. Exploration of the potentials of crosstalk between the Wnt/β-catenin and canonical NF-κB signalling via competitive β-TrCP sequestration

β-TrCP recognises identical phosphorylated amino acid sequence in β-catenin and IκB and mediates their ubiquitination and subsequent degradation by the proteasome (Ougolkov et al., 2004; Spiegelman et al., 2000; Winston et al., 1999). In addition, β-TrCP is generally considered to be low expressed (Fuchs et al., 2004). These two properties of β-TrCP have led to the hypothesis that its substrates may compete for a limited pool of β-TrCP (Fuchs et al., 2004). This chapter explores the conditions that may allow for crosstalk (see Section 2.5) between the Wnt/β-catenin and canonical NF-κB signalling pathways by competitive β-TrCP sequestration into either pathway in a mathematical modelling approach.

Until today, mathematical models are missing that explicitly incorporate β-TrCP dynamics and/or consider interaction of both signalling pathways via β-TrCP. In Section 5.1, a minimal model of competitive β-TrCP sequestration is developed that links two reduced and simplified modules of canonical NF-κB and Wnt/β-catenin signalling via the dynamics of β-TrCP. The minimal model is parameterised in Section 5.2. It is used to identify the conditions that enable or prevent crosstalk between both pathways via competitive β-TrCP sequestration (Section 5.3). The simulations show that the simultaneous stimulation with Wnt hardly affects TNF-stimulus-induced NF-κB dynamics. In contrast, simultaneous stimulation with TNF can influence Wnt/β-catenin signal transduction. The analysis reveals that β-TrCP-mediated IκB degradation, β-TrCP production, and β-TrCP degradation predominantly affect this crosstalk. Low β-TrCP abundance, however, seems unnecessary to enable crosstalk via competitive β-TrCP sequestration.

5.1 Development of a minimal model of competitive β-TrCP sequestration

5.1.1 Structure of the model

The minimal model of competitive β-TrCP sequestration, hereafter referred to as minimal model, is composed of two modules that describe key processes of Wnt/β-catenin signalling and canonical NF-κB signalling pathways (Figure 5.1, red and blue, respectively). The two pathway modules are linked via β-TrCP, which can be sequestered into either of them. Experimental reports have demonstrated that β-TrCP associates with the destruction complex to bind β-catenin and to the NF-κB/IκB complex associated to the IKK complex (Chen et al., 1995; Maniatis, 1999; Tsuchiya et al., 2010).

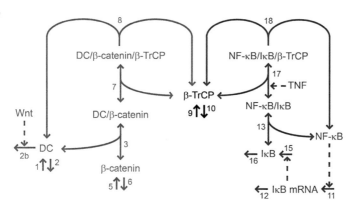

Figure 5.1: Reaction scheme of the minimal model of competitive β-TrCP sequestration.
Schematic representation of the reactions of the minimal model of competitive β-TrCP sequestration. Processes of the Wnt/β-catenin signalling module and the canonical NF-κB signalling module are color-coded in red and blue, respectively. The two modules are linked via β-TrCP (black). The number next to an arrow denotes the number of the particular reaction. One-headed arrows denote reactions taking place in the indicated direction. Double-headed arrows illustrate binding reactions. Dashed arrows represent activation steps. Components in a complex are separated by slashes. DC: destruction complex.

The Wnt/β-catenin signalling module includes four species: β-catenin, destruction complex (DC), DC/β-catenin, and DC/β-catenin/β-TrCP (Figure 5.1, red). The DC is produced (Reaction 1) and degraded in a Wnt stimulus independent as well as dependent manner (Reaction 2 and Reaction 2b, respectively). β-Catenin, produced and degraded in Reaction 5 and Reaction 6, respectively, associates reversibly with the DC to form the

DC/β-catenin complex (Reaction 3). The DC/β-catenin complex reversibly binds to β-TrCP to form the DC/β-catenin/β-TrCP complex (Reaction 7). The DC/β-catenin/β-TrCP complex degrades β-catenin resulting in the release of β-TrCP and the DC (Reaction 8).

In the canonical NF-κB signalling module (Figure 5.1, blue), NF-κB reversibly associates with its inhibitor IκB forming the NF-κB/IκB complex (Reaction 13). This complex reversibly binds to β-TrCP to form the NF-κB/IκB/β-TrCP complex (Reaction 17). In this reaction, the complex formation depends on the presence of a TNF stimulus, while the reverse reaction of complex dissociation occurs independently of a TNF stimulus (Equation 7.169, Appendix 7.4.3). The NF-κB/IκB/β-TrCP complex degrades IκB and releases β-TrCP and NF-κB (Reaction 18). NF-κB induces the expression of IκB mRNA (Reaction 11). IκB mRNA is degraded via Reaction 12. Finally, IκB is translated from IκB mRNA (Reaction 15) and degraded (Reaction 16).

β-TrCP is produced *de novo* (Reaction 9) and degraded (Reaction 10) (Figure 5.1, black). It binds either to the DC/β-catenin complex in the Wnt/β-catenin signalling module (Reaction 7) or to the NF-κB/IκB complex of the canonical NF-κB signalling module (Reaction 17).

Overall, the minimal model of competitive β-TrCP sequestration consists of nine ODEs and one algebraic equation describing the conservation relation of total NF-κB (Equations 7.144 – 7.153, Appendix 7.4).

5.1.2 Kinetics used in the minimal model

Degradation rates, complex formation as well as complex dissociation rates are described by mass action kinetics (Appendix 7.4.3). Generally, production rates are considered to be constant. The two exceptions are IκB mRNA production and IκB translation (Reaction 11 and Reaction 15, respectively, in Figure 5.1). They are assumed to be highly non-linear processes in the cell and are modelled by Hill-type activation kinetics (Equation 7.164 and Equation 7.167, Appendix 7.4.3).

5.2 Parameterisation of the minimal model of competitive β-TrCP sequestration

Until today, there are no experimental time series datasets available concerning β-TrCP as well as components of the canonical NF-κB and Wnt/β-catenin pathway in one cell type. Consequently, kinetic parameters associated with β-TrCP recruitment into the

NF-κB/IκB/β-TrCP complex and the DC/β-catenin/β-TrCP complex in the minimal model cannot be estimated from experimental data. Even the relative ratios of the concentrations of NF-κB, β-catenin, and β-TrCP in one cell type are unknown. All together, the lack of experimental data renders the parameterisation of the minimal model difficult. To address this problem, the information on temporal changes of the concentrations of pathway components provided by the detailed models of canonical NF-κB and Wnt/β-catenin signalling (Section 2.1 and Section 2.2, respectively) is used. These two models are chosen because they have been well validated by experiments and were shown to quantitatively describe the dynamics of many components in these signalling pathways, albeit in different cell types.

The approach is described in more detail in Section 5.2.1 to Section 5.2.4. In Section 5.2.1, the detailed models of canonical NF-κB and Wnt/β-catenin signalling, are used to generate time series data of selected species of both pathways. In Section 5.2.2, the strategies to simulate the Wnt and TNF stimulus in the minimal model are described. Then, the parameter estimation of the Wnt/β-catenin signalling module (Section 5.2.3) and of the canonical NF-κB signalling module (Section 5.2.4) in the minimal model is explained. Finally, the resulting estimated parameter sets are summarised and briefly discussed (Section 5.2.5).

5.2.1 Generation of data sets describing the dynamics of selected pathway components

At first, species of the detailed models of canonical NF-κB and Wnt/β-catenin signalling are selected that are considered to correspond to the species of the minimal model (Table 5.1). In the case of IκB and NF-κB, the sum of the nuclear and cytoplasmic species in the detailed models of canonical NF-κB signalling is considered to correspond to the species IκB and NF-κB of the minimal model, respectively. Note that the concentration of the nuclear species in the sum is weighted by the cytoplasmic-to-nuclear-volume-ratio (k_v, listed in Table 7.1 in Appendix 7.1) to account for the different volumes of both cellular compartments in the detailed model of canonical NF-κB signalling. The sum of nuclear and cytoplasmic IκB as well as the sum of nuclear and cytoplasmic NF-κB are hereafter referred to as total unbound IκB and total unbound NF-κB, respectively.

Table 5.1: List of corresponding species in the different models.

The table lists the species of the minimal model (first column) and the species of the detailed models of Wnt/β-catenin and canonical NF-κB signalling (second and third column, respectively) that are considered to correspond to each other. DC: destruction complex; asterisk denotes phosphorylation; nuc: nuclear species; k_v: parameter of cytosolic-to-nuclear-volume-ratio (see text of Section 5.2.1 for further explanation).

Species of the minimal model (Figure 5.1)	Species of the detailed model of Wnt/β-catenin signalling (Figure 2.2)	Species of the detailed model of canonical NF-κB signalling (Figure 2.1)
(β-catenin)	(β-catenin)	
(DC)	(Axin*/APC*/GSK3)	
(DC/β-catenin)	(Axin*/APC*/GSK3/β-catenin)	
(IκB mRNA)		(IκB-mRNA)
(IκB)		$\frac{1}{k_v} \cdot (\text{IκBnuc}) + (\text{IκB})$
(NF-κB)		$\frac{1}{k_v} \cdot (\text{NF-κBnuc}) + (\text{NF-κB})$

Next, the detailed model of Wnt/β-catenin signalling is simulated to sample a set of time series data points that represent the dynamics of β-catenin, Axin*/APC*/GSK3 complex, and Axin*/APC*/GSK3/β-catenin complex in response to transient Wnt stimulation. Then, the detailed model of canonical NF-κB signalling is simulated to provide a set of data points that capture the dynamics of IκB-mRNA, nuclear and cytoplasmic IκB, and nuclear and cytoplasmic NF-κB in response to TNF stimulation. In all sampled time series data sets, the time points were selected such that characteristic features of the dynamics were represented. These are in particular the unstimulated and stimulated steady states as well as maxima and minima of the transient dynamics. For this reason, the time points are not equally distributed over time. Furthermore, the time points of one pathway component do not necessarily coincide with those of another pathway component (see Figure 5.3).

5.2.2 Modelling the transient Wnt and transient TNF stimulus of the minimal model

In the detailed model of Wnt/β-catenin signalling, the transient Wnt stimulus is given by an exponential decay function (Equation 7.83, Appendix 7.2.5). Accordingly, the transient Wnt stimulus in the minimal model is also described by an exponential decay function (Equation 7.171, Appendix 7.4.5). The half-live of the transient Wnt stimulus (parameter λ in Equation 7.171) in the minimal model is estimated simultaneously with the other parameters of the Wnt/β-catenin signalling module (see Section 5.2.3).

In the detailed model of canonical NF-κB signalling, no transient TNF stimulus is considered. To simulate a TNF stimulus in the detailed model of canonical NF-κB signalling, TNF is incremented from zero to the value of 1 upon stimulation and kept constant over time (Equation 7.44, Appendix 7.1). This constant TNF translates into a transient increase of IKKa concentration in the detailed model. This transient increase of IKKa concentration then induces the association of IKKa and IκB and IKK-dependent phosphorylation of IκB in the detailed model (Reaction 7 and Reaction 9 in Figure 2.1). In order to model a transient TNF stimulation in the minimal model, the IKKa dynamics rather than the constant TNF stimulus of the detailed model of canonical NF-κB signalling are considered. The reactions in the detailed model of canonical NF-κB signalling that contribute to the dynamics of IKKa (Figure 2.1, Reactions 1 – 10, and Reaction 26) form a very complex network. Therefore, they are not feasible to model the transient TNF stimulus in the minimal model. Instead, a simpler algebraic function, which was introduced as the impulse model (Chechik and Koller, 2009), is considered. The impulse model was shown to robustly estimate temporal profiles of various shapes (Chechik and Koller, 2009) by encoding the temporal changes of a species C[t] (i.e., TNF) as a product of two sigmoidal functions (Equation 5.1).

$$C[t] = \frac{1}{h_1} \cdot \left(h_0 + (h_1 - h_0)\frac{1}{1 + Exp[-\beta_1 \cdot (t - t_1)]} \right) \cdot \left(h_2 + (h_1 - h_2)\frac{1}{1 + Exp[\beta_2 \cdot (t - t_2)]} \right) \qquad 5.1$$

An advantage of the impulse model is that its parameters represent biologically meaningful measures of response characteristics: initial (h_0) and final steady state (h_2), peak amplitude (h_1), slope (β_1 and β_2), as well as onset (t_1) and offset time (t_2) of the response dynamics (Chechik and Koller, 2009). The transient TNF stimulus in the minimal model (Equation 7.172, Appendix 7.4.5) is approximated by the impulse model using Equation 5.1. The parameters of Equation 5.1 are estimated by minimising the sum of squares of difference between the transient TNF stimulus of the minimal model (i.e. the impulse model) and the simulated dataset of IKKa dynamics of the detailed model of canonical NF-κB signalling. The estimated parameter set of the best fit is listed in Table 7.11 (Appendix 7.4). For this estimated parameter set, the transient TNF stimulus is compared to the simulated IKKa data points in Figure 5.2 (grey line and black dots, respectively). The comparison shows that both dynamics are very similar (adjusted R^2 of approximately 0.996). Thus, the simulation of the TNF stimulus using the impulse model is a very good estimation of the IKKa dynamics of the detailed model of canonical NF-κB signalling.

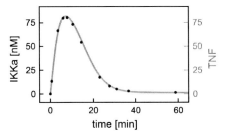

Figure 5.2: The transient TNF stimulus of the minimal model

Comparison between the transient TNF stimulus dynamics of the minimal model (grey line) and the simulated sets of points of the IKKa dynamics of the detailed model of canonical NF-κB signalling (black dots). The transient TNF stimulus retraces the dynamics of IKKa of the detailed model of canonical NF-κB signalling very well. The parameter set is listed in Table 7.11 (Appendix 7.4).

Note that the unstimulated and the stimulated steady state of IKKa differ. While the unstimulated steady state is exactly zero, the stimulated steady state is greater zero (approximately 1.25 nM) establishing a remaining basal activity. These properties of the IKKa dynamics have to be mentioned, because they are preserved in the transient TNF stimulus of the minimal model and of importance to the following analysis. Corresponding to the unstimulated steady state concentration of IKKa, also the initial value of the transient TNF stimulation is zero, i.e. the stimulus is absent. This is of interest because β-TrCP binds NF-κB/IκB only in the presence of TNF stimulation (Equation 7.169, Appendix 7.4) and consequently the unstimulated steady state concentration of the NF-κB/IκB/β-TrCP complex is exactly zero in the case of absent TNF stimulus. If the TNF stimulus and subsequently the NF-κB/IκB/β-TrCP complex concentration are zero then all reactions of the Wnt/β-catenin signalling module as well as those of β-TrCP production and degradation (Figure 5.1, Reactions 1 - 10) are separated from the reactions of the canonical NF-κB signalling module (Reactions 11 - 18).

As mentioned above, the stimulated steady state concentration of IKKa is greater than zero (approximately 1.25 nM) in the detailed model of canonical NF-κB signalling. Similarly, the transient TNF stimulus approaches a value greater than zero (approximately 1.26 nM) after its transient increase in the minimal model (Figure 5.2, grey line), resulting in a remaining basal pathway activity.

The transient Wnt stimulus differs from the transient TNF stimulus in the minimal model, as the transient Wnt stimulus returns to its initial value of zero again after its transient increase. Thus, the Wnt stimulated and Wnt unstimulated steady state of the minimal model are identical.

5.2.3   Estimation of parameters of the Wnt/β-catenin signalling module

So far, the data sets that describe the dynamics of pathway components have been generated in Section 5.2.1 and the transient stimuli of Wnt and TNF have been modelled as described in Section 5.2.2. Next, the minimal model is fitted to the simulated data sets using the simulated annealing algorithm (Kirkpatrick et al., 1983). This algorithm randomly explores the parameter space and tries to minimise the difference between the simulated data sets and the dynamics predicted by the minimal model. The algorithm is allowed to accept solutions that are worse than the current optimal solution, albeit with decreasing frequency during each fitting run. Rather than to find the single parameter set, which yields the globally optimal fit, the aim here is to find a set of distinct parameter sets that sufficiently well reproduce the simulated data sets. Ideally, these different parameter sets cover the whole parameter space. In practice however, only a few hundred of the best fits can be considered for further analysis and simulation (see Section 5.2.5).

In the fitting procedure, only one transient stimulus (either Wnt or TNF) is applied in the minimal model. To begin with, the transient Wnt stimulus in the absence of a TNF stimulus is considered. As discussed in Section 5.2.2, the absence of a TNF stimulus disconnects the reactions of the canonical NF-κB signalling module (Figure 5.1, blue Reactions 11 - 18) from those of the Wnt/β-catenin signalling module as well as those of β-TrCP production and degradation (Reactions 1 - 10). Consequently, the Wnt stimulus does not affect the five species of the canonical NF-κB signalling module (Figure 5.1, blue). The species as well as the reactions of the canonical NF-κB module (Reactions 11 – 18) are therefore neglected in this initial step of the fitting procedure.

To further reduce the parameter space to be explored, four parameter values of the minimal model are adapted from the corresponding parameter of the detailed model of Wnt/β-catenin signalling. These are the parameters v_1, k_2, v_5, and k_6 listed in Table 5.2. The parameter k_{3r} scales to k_3 by the fixed scaling factor set by the parameter K_8 of the detailed model of Wnt/β-catenin signalling (Table 5.2). The remaining parameters k_{2b}, λ, k_3, k_7, k_{7r}, k_8, v_9, and k_{10} are estimated by comparing the dynamics of β-catenin, the DC, and the DC/β-catenin complex of the minimal model to the simulated data sets of β-catenin, Axin*/APC*/GSK3 complex, and Axin*/APC*/GSK3/β-catenin complex of the detailed model, respectively (Table 5.1). To this end, 4400 runs of simulated annealing are performed. The initial values of the parameters k_{2b}, k_3, k_7, k_{7r}, k_8, v_9, and k_{10} were sampled from the interval 10^{-8} to 10^8. The parameter sets yielding the 660 best fits (i.e., best 15% of all 4400 runs) are selected and subsequently used in the estimation of the parameters in the canonical NF-κB signalling module (Section 5.2.4).

Table 5.2: Parameters of the minimal model adapted from the detailed models.

List of the parameters of the minimal model whose values are adapted from that of the detailed models of Wnt/β-catenin or canonical NF-κB signalling. The values of these parameters are kept constant during the fitting procedure.

Parameters of the minimal model (Figure 5.1)	Parameters of the detailed model of Wnt/β-catenin signalling (Figure 2.2)	Parameters of detailed model of canonical NF-κB signalling (Figure 2.1)
v_1	v_{14}	
$\dfrac{v_1}{k_2}$	$(Axin^*/APC^*/GSK3)_{\text{unstimulated steady state}}$	
v_5	v_{12}	
k_6	k_{13}	
$\dfrac{k_{3r}}{k_3}$	K_8	
k_{12}		k_{19}
k_{16}		k_{21}
$(NF\text{-}\kappa B)_{\text{total}}$		$(NF\text{-}\kappa B)_{\text{total}}$

5.2.4 Estimation of parameters in the canonical NF-κB signalling module

In Section 5.2.3, the dynamics of the minimal model upon transient Wnt stimulation in the absence of TNF stimulation were fitted. The procedure yielded 660 different parameter sets that characterise the parameters of Reactions 1 - 10 (Figure 5.1). Next, the minimal model is fitted to the simulated data sets of IκB-mRNA, of total unbound IκB, and of total unbound NF-κB of the detailed model of canonical NF-κB signalling (Table 5.1). This time, a transient TNF stimulus is applied to the minimal model while the Wnt stimulus is absent. If a TNF stimulus is applied the NF-κB signalling module and the Wnt/β-catenin signalling module are linked via the dynamics of β-TrCP. Thus, in contrast to Section 5.2.3, all reactions of the minimal model (Reactions 1 - 18 in Figure 5.1) have to be taken into account.

To reduce the parameter space that has to be explored, parameters $(NF-κB)_{total}$, k_{12}, and k_{16} of the minimal model are set to the corresponding parameter values of total NF-κB, IκB mRNA degradation, and IκB degradation of the detailed model of canonical NF-κB signalling (Table 5.2). The parameters of the Wnt/β-catenin signalling module as well as β-TrCP production and degradation are fixed to one randomly chosen parameter sets from the 660 parameter sets estimated in Section 5.2.3 during each simulated annealing run. The remaining parameters of the canonical NF-κB signalling module were estimated in 3900 independent runs of simulated annealing. Their initial values were generally sampled from the interval 10^{-8} to 10^{8}. In half of the 3900 runs, an additional constrain was imposed on the parameter ratio of k_{17} to k_{17r}. Their ratio was fixed at the value of the particular parameter ratio k_7 to k_{7r} during the run to account for equal IκB-β-TrCP and β-catenin-β-TrCP-binding constants as observed in *in vitro* experiments (Wu et al., 2003). The 400 parameter sets that best fitted the dynamics of IκB mRNA, IκB, and NF-κB of the minimal model to the simulated data sets of the IκB mRNA, total unbound IκB, and total unbound NF-κB, respectively, were selected for further use in this thesis.

5.2.5 Description of the 400 selected parameter sets

In the last section, 400 parameter sets were selected that fit the simulated data sets of the detailed models of canonical NF-κB and Wnt/β-catenin signalling. For all 400 selected parameter sets, simulations of the minimal model are shown in Figure 5.3 (grey lines). To perform these simulations, the minimal model is stimulated with either a transient Wnt

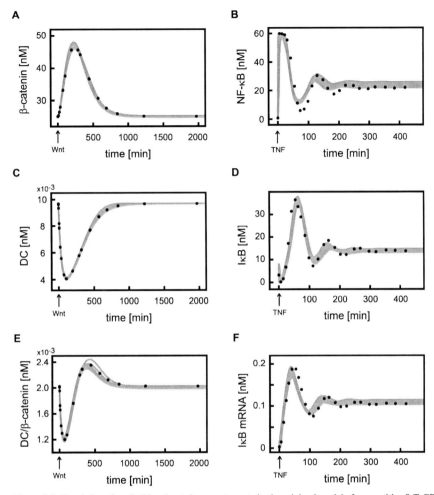

Figure 5.3: Simulations for all 400 estimated parameter sets in the minimal model of competitive β-TrCP sequestration.

The minimal model reproduces the considered dynamics of the detailed models for all 400 selected parameter sets. (A, C, E) Dynamics of species of the Wnt/β-catenin signalling module upon transient Wnt stimulation (grey lines) are compared to the corresponding simulated data sets derived from the detailed model of Wnt/β-catenin signalling (black dots). (B, D, F) Dynamics of species of the canonical NF-κB signalling module of the minimal model upon transient TNF stimulation (grey lines) are compared to the corresponding simulated data sets derived from the detailed model of NF-κB signalling (black dots). Simulations for all 400 selected parameter sets are shown.

stimulus (Figure 5.3A, C, and E) or a transient TNF stimulus (Figure 5.3B, D, and F). In the case of transient Wnt stimulation, the dynamics of β-catenin, the DC, and the DC/β-catenin complex of the minimal model are compared to the simulated data sets of β-catenin, Axin*/APC*/GSK3 complex, and Axin*/APC*/GSK3/β-catenin complex, respectively (Figure 5.3A, C, and E). The comparison shows that the simulated data sets are very well reproduced by the minimal model for all 400 selected parameter sets. To quantify this visual comparison, the coefficient of determination (adjusted R^2) is calculated for all 400 parameter sets. The coefficient of determination indicates how well the dynamics of the minimal model fit the simulated data sets of the detailed models of canonical NF-κB and Wnt/β-catenin signalling. A coefficient of determination of 1 indicates perfect agreement. The median adjusted R^2 of the dynamics of β-catenin, the DC, and the DC/β-catenin complex for all 400 selected parameter sets is 0.99, indicating strong agreement between the dynamics of the minimal model and the simulated data sets on average. In the case of transient TNF stimulation, the dynamics of IκB mRNA, IκB, and NF-κB are compared to the simulated data sets of IκB mRNA, total unbound IκB, and total unbound NF-κB, respectively (Figure 5.3B, D, and F). The simulations show also in this case that the minimal model sufficiently approximates the simulated data sets of the detailed NF-κB model. The median adjusted R^2 of the dynamics of IκB mRNA, IκB, and NF-κB or all 400 selected parameter sets is 0.90. Taken together, the analysis shows that all 400 selected parameter sets of the minimal model reproduce the considered dynamics of the detailed models.

Next, the distribution of the parameter values of the 400 selected parameter sets is examined (Figure 5.4). For each parameter, the median value (thick black line), the range from 25% to 75% quantile (grey box), and the maximal and minimal values ("T" and inverted "T", respectively) are indicated in the box plot. The parameters v_1, k_2, v_5, k_6, k_{12}, and k_{16} are only represented by the median value, because they obtain only one specific value each in all 400 selected parameter sets. Their values were adapted from the detailed models of Wnt/β-catenin and canonical NF-κB signalling (Table 5.2) and not changed during the fitting procedure. Similarly, the ratio of k_{3r} to k_3 was fixed to the parameter K_8 of the detailed model of Wnt/β-catenin signalling (Table 5.2). The values of k_3 and k_{3r} cover about four orders of magnitude (Figure 5.4). The parameters k_{2b}, λ, $vmax_{11}$, K_{11}, k_{13}, $vmax_{15}$, and K_{15} could be well estimated. They have narrow distributions of their values spanning one to two orders of magnitude (Figure 5.4).

In contrast, the parameters k_7, k_{7r}, v_9, k_{10}, k_{17}, and k_{17r} have very broad distributions of their values covering more than ten orders of magnitude (Figure 5.4). These parameters are associated with reactions that involve β-TrCP in the minimal model (Reaction 7, Reaction 9, Reaction 10, and Reaction 17 in Figure 5.1). The finding that the parameter values still cover approximately the initial sampling interval of the fitting procedure (10^{-8} to 10^8) and have not been restricted to a narrow range in the fitting process may be due to the lack of data points describing the dynamics of β-TrCP.

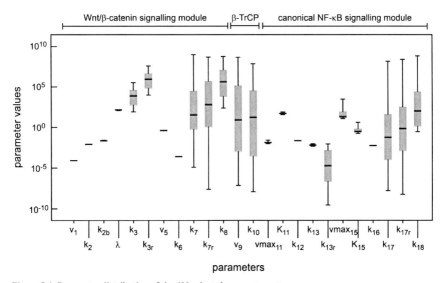

Figure 5.4: Parameter distribution of the 400 selected parameter sets.

The box plot shows the distribution of each parameter in the 400 selected parameter sets. The median (thick black line), the range from 25% to 75% quantile (grey box), and the maximal and minimal values ("T" and inverted "T", respectively) are indicated. Index of parameters refers to the corresponding reaction number. Parameters are grouped on top of the plot according to their respective modules in Figure 5.1.

The parameters k_{13r}, as well as k_8 and k_{18} seem to be restricted on one side by a specific threshold value, below which (k_{13r}) or above which (k_8 and k_{18}) they may obtain any possible value in the considered interval of 10^{-8} to 10^8. In particular, the parameter k_{13r} is restricted to values below 1 min^{-1} (Figure 5.4). This range agrees with the range of parameter values ($0 - 0.06 \text{ min}^{-1}$) used by published mathematical models of NF-κB signalling (Hoffmann et al., 2002; Lipniacki et al., 2004). In the case of parameter k_{18}, the values range approximately from $1\,\text{min}^{-1}$ to $10^8\,\text{min}^{-1}$ (Figure 5.4). Note that a similar range was already discussed for the corresponding parameter in the detailed model of canonical NF-κB signalling (Section 3.2.2). There, it was shown that any value of the rate constant of IKK-dependent IκB degradation ($k_{\beta\text{-TrCP}}$) that is greater $1\,\text{min}^{-1}$ hardly influences the model dynamics. It might thus be speculated that the range of parameter k_{18} of the minimal model cannot be further restricted because any value of parameter k_{18} above $1\,\text{min}^{-1}$ would produce similar dynamics. Similar to parameter k_{18}, the values of parameter k_8 also seem to have a lower bound. The parameter values range from approximately $10^2\,\text{min}^{-1}$ to $10^8\,\text{min}^{-1}$ (Figure 5.4). This lower bound of the parameter range is in agreement with the corresponding parameter value of the detailed model of Wnt/β-catenin signalling. There, the dissociation rate of ubiquitinated β-catenin from the destruction complex (k_{10}) is $2.06 \cdot 10^2\,\text{min}^{-1}$ (Table 7.3, Appendix 7.2). Simulations of the detailed model of Wnt/β-catenin signalling using increased values of parameter k_{10} furthermore revealed that any larger value has no observable impact on β-catenin dynamics in the model (not shown).

Taken together, the minimal model reproduces the considered dynamics of the detailed models well using the 400 selected parameter sets. Seven parameters were restricted by the fitting process to narrow distributions of one to two orders of magnitude. However, the parameters k_7, k_{7r}, v_9, k_{10}, k_{17}, and k_{17r} that are associated with reactions that involve β-TrCP (Reaction 7, Reaction 9, Reaction 10, and Reaction 17 in Figure 5.1) however cover a very broad range of parameter values (more than ten orders of magnitude).

In the analyses shown in Section 5.3, the impact of the parameters on potential crosstalk between Wnt/β-catenin and canonical NF-κB signalling via competitive β-TrCP sequestration is investigated. In general, all 400 selected parameter sets resulting from the fitting procedure are used in the simulations and analyses. The only exception is in Section 5.3.4. There, only one parameter set, which was randomly chosen from the 400 selected parameter sets, is considered in the simulations for sake of clarity. This parameter set is hereafter referred to as reference parameter set and it is given in Table 7.10 and Table 7.11 (Appendix 7.4.4). In the analyses of Section 5.3.2 and Section 5.3.6, changes in the parameter values are applied to all 400 selected parameter sets. In Section 5.3.2, the impact of changes in β-TrCP abundance on the stimulated steady state concentrations of β-catenin and NF-κB are investigated. To that end, the parameter values of v_9 of all 400 selected parameter sets are multiplied by different scaling factors f_{scale}. In Section 5.3.6, the effects of the reduction of production and degradation rates of β-TrCP on the impact of TNF stimulation on Wnt/β-catenin signalling is analysed. In that case, the parameters v_9 and k_{10} were simultaneously changed in all of the 400 selected parameter sets. Further details are provided in Section 5.3.2 and Section 5.3.6.

5.3   Analysis of the minimal model of competitive β-TrCP sequestration

In Section 5.2, parameter estimation of the Wnt/β-catenin and canonical NF-κB signalling module of the minimal model was performed using either pathway-specific Wnt or TNF stimulation, respectively. Hereafter, the dynamics of NF-κB and β-catenin upon simultaneous stimulation with Wnt and TNF stimulus is compared to pathway-specific stimulation. In this way, the influence of the pathway-extrinsic stimulus on the pathway-specific signal transduction is investigated. The analyses focus on the steady state (Sections 5.3.1 - 5.3.3) as well as the transient dynamics of β-catenin and NF-κB upon stimulation with Wnt and TNF (Sections 5.3.4 - 5.3.6).

5.3.1   Steady states of β-catenin and NF-κB do not influence each other

The steady state concentrations of all species of the minimal model can be calculated analytically (Appendix 9.1). The steady state concentration of β-TrCP (Equation 9.11, Appendix 9.1) is given by the ratio of its production and degradation rate constants (v_9 and k_{10}, respectively). It is independent of any Wnt and TNF stimulus.

The steady state equation of β-catenin (Equation 9.17) reveals that the unstimulated steady state concentration and the stimulated steady state concentration are identical. They depend on the parameters of the Wnt/β-catenin signalling module (v_1, k_2, k_3, k_{3r}, v_5, k_6, k_7, k_{7r}, and k_8) as well as the steady state concentration of β-TrCP (ratio of v_9 and k_{10}). The steady state concentrations of β-catenin do not depend on any of the parameters of the canonical NF-κB signalling module.

The steady state equation of NF-κB (Equation 9.23) reveals that the unstimulated and stimulated steady state concentrations depend on the parameters of the canonical NF-κB signalling module ($vmax_{11}$, K_{11}, k_{12}, k_{13}, k_{13r}, $vmax_{15}$, K_{15}, k_{16}, k_{17}, k_{17r}, k_{18}, and NF-κB$_{total}$), the TNF stimulus (TNF$_{stst}$), and also the steady state concentration of β-TrCP (ratio of v_9 and k_{10}). The unstimulated and stimulated steady state concentrations do not depend on the parameters of the Wnt/β-catenin signalling module. The dependence of the steady state concentrations of NF-κB on the TNF stimulus (TNF$_{stst}$) means that the TNF stimulated and unstimulated steady state concentration differ, because the initial value (unstimulated TNF_{stst}) and final value (stimulated TNF_{stst}) of the transient TNF stimulation differ (see Section 5.2.2). Taken together, the steady state equations demonstrate that stimulated and unstimulated steady state concentrations of β-catenin and NF-κB only depend on the parameters of the respective signalling module and the steady state concentration of β-TrCP. Consequently, the Wnt/β-catenin signalling module and the canonical NF-κB signalling module do not influence each other at unstimulated and stimulated steady state conditions in the minimal model. This observation allows considering merely the steady state upon simultaneous stimulation with Wnt and TNF in the following analyses (Section 5.3.2 and Section 5.3.3), since identical results will be obtained for β-catenin and NF-κB in case of single Wnt or TNF stimulation, respectively. For sake of convenience, the steady state upon simultaneous stimulation with Wnt and TNF will be referred to as stimulated steady state in Section 5.3.2 and Section 5.3.3.

5.3.2 Modulation of β-TrCP abundance regulates the stimulated steady state concentrations of β-catenin and NF-κB in opposite directions

The steady state equations of β-catenin and NF-κB (Equation 9.17 and Equation 9.23, respectively, in Appendix 9.1) revealed that the stimulated steady states of β-catenin and NF-κB depend on exclusive disjunct sets of parameters, with the exception that both steady states depend on the ratio of production rate to degradation rate constant of β-TrCP (Section 5.3.1). The ratio of production rate to degradation rate constant of β-TrCP defines the steady state concentration of β-TrCP (Equation 9.11, Appendix 9.1). In this section, the influence of β-TrCP abundance on the stimulated steady states of β-catenin and NF-κB is investigated. To that end, the stimulated steady states of NF-κB and β-catenin are calculated for different steady state concentrations of β-TrCP.

At first, the steady state concentrations of β-TrCP are calculated for the 400 selected parameter sets. These 400 steady state concentrations of β-TrCP span over 14 orders of magnitude ranging from 10^{-7} nM to 10^{7} nM (Figure 5.5A). Next, each of these 400 steady state concentrations of β-TrCP is subsequently varied by changing the respective parameter v_9 by a scaling factor f_{scale} ranging from 10^{-5} to 10^{3}. In case of $f_{scale} < 1$, the value of parameter v_9 is reduced resulting in the down-regulation of β-TrCP abundance. In case of $f_{scale} > 1$, the value of parameter v_9 increases and β-TrCP abundance is up-regulated. Again, the steady state concentrations of β-catenin and NF-κB are calculated using Equation 9.17 and Equation 9.23 (Appendix 9.1), respectively, and plotted in Figure 5.5B (black and grey lines, respectively).

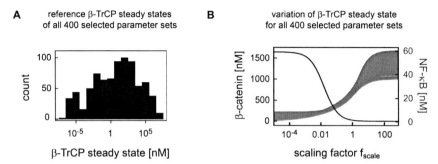

A   reference β-TrCP steady states of all 400 selected parameter sets

B   variation of β-TrCP steady state for all 400 selected parameter sets

Figure 5.5: Impact of β-TrCP abundance on the stimulated steady states of β-catenin and NF-κB.
A modulation of β-TrCP abundance regulates the stimulated steady state concentrations of β-catenin and NF-κB in opposite directions in all 400 selected parameter sets, independent of the original β-TrCP steady state concentration in each parameter set. (A) The histogram shows the distribution of the steady state of β-TrCP calculated for all 400 selected parameter sets. (B) For each of the 400 selected parameter sets, the steady state of β-TrCP steady state, shown in (A), is varied by a scaling factor f_{scale} ranging from 10^{-5} to 10^3. The corresponding stimulated steady states of β-catenin (black lines) and NF-κB (grey lines) are calculated. Note that the 400 curves corresponding to the stimulated steady states of β-catenin lay on top of each other.

The analysis shows that a change in β-TrCP abundance regulates the stimulated steady state concentrations of NF-κB and β-catenin in opposite directions: up-regulation of β-TrCP abundance increases NF-κB and decreases β-catenin concentration, while down-regulation of β-TrCP abundance decreases NF-κB and increases β-catenin concentration (Figure 5.5B). Down-regulation of β-TrCP abundance inhibits the degradation of β-catenin via the destruction complex leading to an accumulation of β-catenin. Up-regulation of β-TrCP, on the other hand, reduces the stimulated steady state concentration of β-catenin due to enhanced degradation of β-catenin via the destruction complex. In contrast to β-catenin, the stimulated steady state concentration of NF-κB decreases if β-TrCP abundance is down-regulated. Down-regulation of β-TrCP inhibits the degradation of the NF-κB inhibitor IκB. Thus, IκB is able to retain most of the NF-κB in the transcriptional inactive NF-κB/IκB complex and the stimulated steady state concentration of unbound NF-κB remains low. Up-regulation of β-TrCP abundance, on the other hand, enhances IκB degradation resulting in elevated stimulated steady state concentration of NF-κB (Figure 5.5B). Note that the up-regulation or down-regulation of β-TrCP causes for all 400 selected parameter sets similar effects on the steady state concentrations of β-catenin and NF-κB (Figure 5.5B), despite the broad distribution of the actual β-TrCP steady state concentration in all 400 selected parameter sets (Figure 5.5A).

5.3.3 Impact of parameter values on the stimulated steady state concentration ranges of NF-κB and β-catenin

A further examination of Figure 5.5B shows that the stimulated steady state concentrations of β-catenin of all 400 simulations lay on top of each other (black lines), while the stimulated steady state concentrations of NF-κB vary evidently between the 400 simulations (grey lines). The variations are most prominent at the smallest and largest value of the scaling factor f_{scale} (10^{-5} and 10^{3}, respectively). The maximal and minimal possible stimulated steady states of β-catenin and NF-κB are found in the limits of the scaling factor f_{scale} approaching zero and infinity, respectively (not shown). These two extreme cases are considered in this section to explore the impact of the parameter values on the maximal and minimal possible stimulated steady states of β-catenin and NF-κB. Note that in the first case, if the scaling factor f_{scale} approaches zero consequently also the steady state of β-TrCP approaches zero for all 400 selected parameter sets. In the second case, if the scaling factor f_{scale} approaches infinity also the steady state of β-TrCP approaches infinity for all 400 selected parameter sets.

Examination of the minimal and maximal possible stimulated steady states of β-catenin

In the limit of β-TrCP steady state concentration approaching zero (i.e. $\frac{v_9}{k_{10}} \to 0$), the stimulated steady state concentration of β-catenin (Equation 9.17, Appendix 9.1) is given by the ratio of β-catenin production (v_5) to degradation rate constant (k_6) (Equation 5.2).

$$(\beta\text{-}catenin)_{stst} = \frac{v_5}{k_6}$$

5.2

Since both parameters v_5 and k_6 have fixed values in all 400 selected parameter sets (Figure 5.4), the stimulated steady state concentrations of β-catenin are identical for all 400 simulations in the limit of β-TrCP concentration approaching zero. This maximal possible steady state concentration of β-catenin is 1645.91 nM. It is in quantitative agreement with the maximal possible β-catenin concentration in the detailed model of Wnt/β-catenin signalling (Section 2.2).

In the limit of β-TrCP concentration approaching infinity (i.e. $\frac{v_9}{k_{10}} \to \infty$), the stimulated steady state concentration of β-catenin (Equation 9.17, Appendix 9.1) can be reduced to Equation 5.3 using l'Hôpital's rule[4].

$$(\beta\text{-}catenin)_{stst} = \frac{v_5}{k_6} \cdot \frac{1}{\left(\frac{v_1}{k_2} \cdot \frac{k_3}{k_6} + 1\right)}$$

5.3

The parameters v_1, k_2, v_5, and k_6 are fixed in all 400 selected parameter sets. Thus, only the parameter k_3 in Equation 5.3 differs between all 400 selected parameter sets. This variation in k_3 results in the variation of the stimulated steady state concentration of β-catenin in the case of very high β-TrCP concentrations. However, the stimulated steady state concentrations of β-catenin for all 400 selected parameter sets are all close to zero (ranging from approximately 1.2 nM to 0.53 nM) such that the variation is hardly detectable in Figure 5.5B.

Altogether, the analysis showed that the maximal possible steady state concentration of β-catenin is fixed at 1645.91 nM for all 400 selected parameter sets. The variation of the minimal possible steady state concentration of β-catenin is created only by the parameter k_3 since the other parameters are fixed in all 400 selected parameter sets.

Examination of the minimal and maximal possible stimulated steady states of NF-κB

In contrast to the stimulated steady state concentrations of β-catenin, the minimal and maximal stimulated steady state concentrations of NF-κB vary evidently comparing the 400 simulations in Figure 5.5B (grey lines). The possible minimal concentrations in the minimal model range from approximately 1 nM to 8 nM. The possible maximal concentrations span from about 37 nM to 60 nM. In both cases, the concentration ranges are greater than the

---

[4] l'Hôpital's rule in its simplest form:

If $\lim_{x \to c} f(x) = \lim_{x \to c} g(x) = 0$ or $\lim_{x \to c} f(x) = \lim_{x \to c} g(x) = \pm\infty$, and $\lim_{x \to c} \frac{f'(x)}{g'(x)}$ exists, and $g'(x) \neq 0$ for all x, then $\lim_{x \to c} \frac{f(x)}{g(x)} = \lim_{x \to c} \frac{f'(x)}{g'(x)}$.

concentration obtained for total unbound NF-κB in the detailed model of canonical NF-κB

signalling (approximately 0.28 nM and 21 nM, respectively). A possible explanation is that in both models identical total NF-κB concentrations are considered (Table 5.2) but the total NF-κB concentration is distributed over a reduced number of species in the minimal model compared to the detailed kinetic model of canonical NF-κB signalling (Equation 7.153 and Equation 7.15, respectively, in Appendix 7). Consequently, NF-κB concentrations in the minimal model can augment the concentrations of total unbound NF-κB in the detailed kinetic model of canonical NF-κB signalling.

In the limit of β-TrCP concentration approaching zero (i.e. $\frac{v_9}{k_{10}} \to 0$), the polynomial in the stimulated steady state concentration of NF-κB (NF-κB)$_{stst}$ given by Equation 9.23 (Appendix 9.1) can be reduced to Equation 5.4:

$$0 = \frac{k_{13r}}{k_{13}} \cdot ((NF\text{-}\kappa B)_{total} - (NF\text{-}\kappa B)_{stst})$$

$$-\frac{vmax_{15}}{k_{16}} \cdot \frac{\dfrac{((NF\text{-}\kappa B)_{stst})^4}{\left(K_{11}{}^2 + ((NF\text{-}\kappa B)_{stst})^2\right)^2}}{\left(\dfrac{k_{12} \cdot K_{15}}{vmax_{11}}\right)^2 + \dfrac{((NF\text{-}\kappa B)_{stst})^4}{\left(K_{11}{}^2 + ((NF\text{-}\kappa B)_{stst})^2\right)^2}} \cdot (NF\text{-}\kappa B)_{stst} \qquad 5.4$$

In this limit, the parameters vmax$_{11}$, K$_{11}$, k$_{12}$, k$_{13}$, k$_{13r}$, vmax$_{15}$, K$_{15}$, k$_{16}$, and (NF-κB)$_{total}$ determine the minimal possible stimulated steady state concentrations of NF-κB for the 400 selected parameter sets. The parameters k$_{12}$ and k$_{16}$ are fixed in all 400 selected parameter sets and thus do not contribute to the observed variations. The parameters k$_{17}$, k$_{17r}$, and k$_{18}$ have no influence on the steady state concentrations of NF-κB if β-TrCP concentrations approach zero, because all summands of Equation 9.23 (Appendix 9.1) that contain these parameters become zero. In the same way, the steady state becomes independent of the TNF stimulus (TNF$_{stst}$). Consequently, the stimulated and the unstimulated steady state concentrations of NF-κB are identical.

In the limit of β-TrCP concentration approaching infinity (i.e. $\frac{v_9}{k_{10}} \to \infty$), the polynomial of the stimulated steady state concentration of NF-κB (NF-κB)$_{stst}$ given by Equation 9.23 (Appendix 9.1) simplifies to Equation 5.5:

$$0 = \frac{k_{18}}{k_{13}} \cdot ((NF\text{-}\kappa B)_{total} - (NF\text{-}\kappa B)_{stst})$$

$$-\frac{vmax_{15}}{k_{16}} \cdot \frac{\dfrac{((NF\text{-}\kappa B)_{stst})^4}{\left(K_{11}{}^2 + ((NF\text{-}\kappa B)_{stst})^2\right)^2}}{\left(\dfrac{k_{12} \cdot K_{15}}{vmax_{11}}\right)^2 + \dfrac{((NF\text{-}\kappa B)_{stst})^4}{\left(K_{11}{}^2 + ((NF\text{-}\kappa B)_{stst})^2\right)^2}} \cdot (NF\text{-}\kappa B)_{stst}$$

$$+\frac{k_{18}}{k_{16}} \cdot ((NF\text{-}\kappa B)_{total} - (NF\text{-}\kappa B)_{stst}) \cdot (NF\text{-}\kappa B)_{stst} \qquad\qquad 5.5$$

Equation 5.5 shows that the parameters $vmax_{11}$, K_{11}, k_{12}, k_{13}, $vmax_{15}$, K_{15}, k_{16}, and (NF-κB)$_{total}$ also determine the maximal possible stimulated steady state concentrations of NF-κB. In addition, the parameter k_{18} has an effect. The parameters k_{17}, k_{17r}, and k_{13r} have no influence on the steady state concentrations of NF-κB if β-TrCP concentrations approach infinity. Similar to the minimal possible steady state concentrations of NF-κB (Equation 5.4), the maximal possible steady state is independent of the TNF stimulus (Equation 5.5).

Taken together, the polynomials for the minimal and maximal possible stimulated steady state concentrations of NF-κB (Equation 5.4 and Equation 5.5, respectively) were examined in the limits of β-TrCP approaching zero and infinity, respectively. The analysis revealed what particular parameters have an influence on the steady state concentration of NF-κB. In contrast to the steady state concentration of β-catenin, where only one parameter (k_3) varied while the others remained fixed (see page 106), eight parameters vary between the 400 selected parameter sets in the case of NF-κB. Next, the impact of the values of these parameters on the variations in the minimal and maximal possible stimulated steady state concentrations of NF-κB (Figure 5.5B) is investigated for the 400 selected parameter sets. The steady state equations of NF-κB (Equation 5.4 and Equation 5.5) show that certain parameters appear in association with always the same other parameters. Thus, they can be grouped into the parameter combinations $\frac{k_{18}}{k_{16}}$, $\frac{k_{18}}{k_{13}}$, $\frac{k_{12} \cdot K_{15}}{vmax_{11}}$, $\frac{vmax_{15}}{k_{16}}$, and $\frac{k_{13r}}{k_{13}}$ as well as the remaining single parameter K_{11} (which is also termed a "parameter combination" in this section for sake of convenience). To quantify the impact of these parameter combinations on the variations of the minimal and maximal possible stimulated steady states of NF-κB, Kendall rank correlation coefficients (Section 2.8) are calculated. The Kendall rank correlation coefficient has been used as a non-parametric measure of association between two sets of observations on the basis of their ranking. The coefficient takes values in the range from -1 to 1; with 1 corresponding to perfect agreement in the two rankings, -1 corresponding to perfect disagreement in the two rankings. If both sets are independent of each other, the Kendall rank correlation coefficient is zero.

In the limit of β-TrCP concentration approaching zero, the Kendall rank correlation coefficients of the parameter combinations are determined with respect to the minimal possible stimulated steady state concentrations of NF-κB (Figure 5.6A). The strongest correlation is found between the parameter combination $\frac{k_{13r}}{k_{13}}$ and the minimal possible stimulated steady state concentration of NF-κB. The Kendall rank correlation coefficient of $\frac{k_{13r}}{k_{13}}$ is approximately 0.95 (Table 9.1, Appendix 9.4). Furthermore, the parameters related to the transcriptional feedback via IκB (K_{11}, $\frac{k_{12} \cdot K_{15}}{vmax_{11}}$, and $\frac{vmax_{15}}{k_{16}}$) correlate (Table 9.1, Appendix 9.4) with the minimal possible stimulated steady state concentration of NF-κB although their Kendall rank correlation coefficients are about 6-fold smaller than that of $\frac{k_{13r}}{k_{13}}$ (Table 9.1, Appendix 9.4). As expected from the analysis of Equation 5.4, the parameter combinations $\frac{k_{18}}{k_{16}}$ and $\frac{k_{18}}{k_{13}}$ have hardly a correlation with the minimal possible stimulated steady state concentration of NF-κB (Table 9.1, Appendix 9.4). The respective Kendall rank correlation coefficients are close to but not exact zero (Table 9.1, Appendix 9.4), which was not expected from the analysis of Equation 5.4. The slight difference of the Kendall rank correlation coefficients from zero is due to a by chance unequal number of concordant and discordant pairings (see Section 2.8) in the 400 selected parameter sets.

The large Kendall rank correlation coefficient of $\frac{k_{13r}}{k_{13}}$ can be explained as follows. The strong down-regulation of β-TrCP to zero inhibits the formation of the NF-κB/IκB/β-TrCP complex despite the presence of a TNF stimulus. Consequently, most of the total NF-κB is distributed between the NF-κB/IκB complex and unbound NF-κB in the minimal model. At steady state, the NF-κB/IκB complex and unbound NF-κB are linked via the dissociation constant of the NF-κB/IκB complex ($\frac{k_{13r}}{k_{13}}$). The dissociation constant of the NF-κB/IκB complex ($\frac{k_{13r}}{k_{13}}$)

influences how the total NF-κB is distributed between the two model species. That explains the strong correlation between $\frac{k_{13r}}{k_{13}}$ and the stimulated steady state concentration of NF-κB.

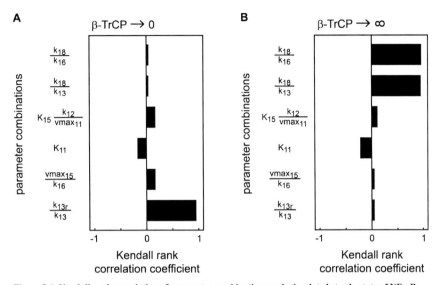

Figure 5.6: Kendall rank correlation of parameter combinations and stimulated steady state of NF-κB.

Distinct parameter combinations correlate with the minimal and maximal possible stimulated steady state of NF-κB. (A) Kendall rank correlation coefficients between parameter combinations and minimal possible stimulated steady state of NF-κB for all 400 selected parameter sets in the case of β-TrCP concentration approaching zero are shown. All Kendall rank correlation coefficients and their corresponding p-values are specified in Table 9.1 (Appendix 9.4). (B) Kendall rank correlation coefficients between parameter combinations and maximal possible stimulated steady state of NF-κB for all 400 selected parameter sets in the case of β-TrCP concentration approaching infinity are shown. All Kendall rank correlation coefficients and their corresponding p-values are specified in Table 9.2 (Appendix 9.4).

In the limit of β-TrCP concentration approaching infinity, the Kendall rank correlation coefficients of the parameter combinations is calculated with respect to the maximal possible stimulated steady states of NF-κB (Figure 5.6B). The Kendall rank correlation coefficients and their corresponding p-values are specified in Table 9.2 (Appendix 9.4). In the limit of β-TrCP concentration approaching infinity, the parameter combinations $\frac{k_{18}}{k_{16}}$ and $\frac{k_{18}}{k_{13}}$ most strongly correlate with the maximal possible stimulated steady state concentration of NF-κB

(Figure 5.6B). The parameters $\frac{k_{18}}{k_{16}}$ and $\frac{k_{18}}{k_{13}}$ have very similar Kendall rank correlation coefficients (0.97 and 0.95, respectively; Table 9.2 in Appendix 9.4). The reason is that parameter k_{16} is fixed and parameter k_{13} hardly varies in all 400 selected parameter sets (Figure 5.4), such that both parameters are merely scaling factors of k_{18} in this analysis. The large Kendall rank correlation coefficients of the parameter combinations related to k_{18} ($\frac{k_{18}}{k_{16}}$ and $\frac{k_{18}}{k_{13}}$) may be explained in the following way. Due to the strong up-regulation of β-TrCP to infinity, almost all available NF-κB/IκB complexes are bound by β-TrCP to form NF-κB/IκB/β-TrCP complexes via Reaction 17 (Figure 5.1). Consequently, total NF-κB is distributed between the NF-κB/IκB/β-TrCP complexes and unbound NF-κB at stimulated steady state. The dissociation reaction of the NF-κB/IκB/β-TrCP complex (Reaction 18 in Figure 5.1) strongly impacts this distribution of total NF-κB, which is mirrored by the large Kendall rank correlation coefficients of the parameter combinations related to k_{18}.

In contrast to the parameter combinations $\frac{k_{18}}{k_{16}}$ and $\frac{k_{18}}{k_{13}}$, the Kendall rank correlation coefficient of the dissociation constant of the NF-κB/IκB complex ($\frac{k_{13r}}{k_{13}}$) is almost zero (Table 9.2, Appendix 9.4). The Kendall rank correlation coefficient is not exact zero, as was expected from the analysis of Equation 5.5, due to a by chance unequal number of concordant and discordant pairings (see Section 2.8) in the 400 selected parameter sets. Similar to the analysis of the minimal possible stimulated steady state concentration of NF-κB, the parameter combinations K_{11} and $\frac{K_{15} \cdot k_{12}}{vmax_{11}}$ also show a correlation with the maximal possible stimulated steady state concentration of NF-κB, although their Kendall rank correlation coefficients are about 9-fold and 4-fold smaller than that of the parameter combinations $\frac{k_{18}}{k_{16}}$ and $\frac{k_{18}}{k_{13}}$ (Table 9.2, Appendix 9.4). The parameter combination $\frac{vmax_{15}}{k_{16}}$ hardly correlates with the maximal possible stimulated steady state concentration of NF-κB (Table 9.2, Appendix 9.4).

5.3.4 TNF stimulation can influence Wnt/β-catenin signal transduction

Having analysed the steady states in Sections 5.3.1 - 5.3.3, Sections 5.3.4 - 5.3.6 focus on the dynamics of β-catenin and NF-κB. First, the influence of a pathway-extrinsic stimulus on the pathway-intrinsic output is investigated. In particular, whether simultaneous stimulation with TNF influences the β-catenin dynamics in response to Wnt stimulation and whether simultaneous stimulation with Wnt modulates the NF-κB dynamics in response to TNF stimulation. To do so, the dynamics upon pathway-specific stimulation are compared with the dynamics upon simultaneous stimulation with pathway-specific and pathway-extrinsic stimuli (Figure 5.7). Figure 5.7A and B show the transient dynamics of β-catenin and NF-κB, respectively, using the reference parameter set (listed in Appendix 7.4.4, Table 7.10 and Table 7.11). While the different stimulations result in distinct β-catenin dynamics, the NF-κB dynamics for both stimulations do not show a difference. Note that β-catenin has not yet reached its stimulated steady state concentration until 1600 min in Figure 5.7A. Considering a longer time range, the full response dynamics are shown in Figure 9.1 (Appendix 9.5).

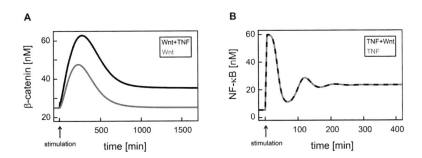

Figure 5.7: Effects of simultaneous TNF and Wnt stimulation on the dynamics of β-catenin and NF-κB, respectively.
Simultaneous stimulation with Wnt and TNF strongly changes the β-catenin dynamics but hardly affects the NF-κB dynamics compared to the respective single pathway-specific stimulations. (A) Dynamics of β-catenin upon Wnt stimulation (grey line) is compared to dynamics of β-catenin upon simultaneous stimulation with Wnt and TNF (black line). Note that β-catenin has not yet reached its stimulated steady state concentration. See Figure 9.1 in Appendix 9.5 for the response dynamics over a longer time range. (B) Dynamics of NF-κB upon TNF stimulation (grey dashed line) is compared to dynamics of NF-κB upon simultaneous stimulation with TNF and Wnt (black dashed line). Note that the two curves lie on top of each other in the plot. All simulations consider the reference parameter set (Table 7.10 and Table 7.11, Appendix 7.4.4).

The analysis of the β-catenin dynamics (Figure 5.7A) reveals that the stimulation with TNF in addition to Wnt increases the signalling time, signal duration, and signal amplitude (Section 2.4) of the β-catenin dynamics by a factor of about 1.9, 5.9, and 1.7, respectively, compared to single Wnt stimulation. That indicates that the concentration of β-catenin rises to a higher level (Figure 5.7A) and remains at increased levels for a longer time period (Figure 9.1, Appendix 9.5) if a TNF stimulus is applied in combination with a Wnt stimulus compared to the single Wnt stimulation. In contrast, simultaneous stimulation with Wnt and TNF does not result in an observable change in the NF-κB dynamics compared to single TNF stimulation (Figure 5.7B).

To further investigate this crosstalk, the dynamics of unbound β-TrCP upon simultaneous Wnt and TNF stimulation is examined. The concentration of unbound β-TrCP decreases strongly in the first minute (Figure 5.8A, black line) due to its sequestration into NF-κB/IκB/β-TrCP complexes (Figure 5.8A, blue line). It increases again in the next 5 min predominantly as a result of NF-κB/IκB/β-TrCP complex dissociation, which is indicated by the decrease of NF-κB/IκB/β-TrCP complex concentration. About 30 min after simultaneous Wnt and TNF stimulation, unbound β-TrCP reaches a concentration of approximately 0.011 nM that is maintained for the next 1500 min. The simulation reveals that already 7 min after simultaneous Wnt and TNF stimulation unbound β-TrCP constitutes at least 60% of the

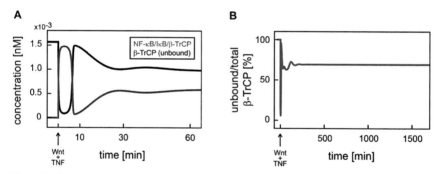

Figure 5.8: Dynamics of unbound β-TrCP.
The high absolut and relative abundance of β-TrCP available for Wnt/β-catenin signalling seems not compatible with the observation of crosstalk in Figure 5.7A (see text of Section 5.3.4 for detailed discussion). (A) Dynamics of unbound β-TrCP (black line) and the NF-κB/IκB/β-TrCP complex (blue line) upon simultaneous stimulation with TNF and Wnt. (B) Dynamics of unbound β-TrCP as fraction of total β-TrCP upon simultaneous stimulation with TNF and Wnt. Total β-TrCP is given by the sum of the concentrations of β-TrCP, NF-κB/IκB/β-TrCP complex, and DC/β-catenin/β-TrCP complex in the minimal model. The reference parameter set (Table 7.10 and Table 7.11, Appendix 7.4.4) is used in the simulations.

total β-TrCP concentration (Figure 5.8B). Thus, enough unbound β-TrCP is available for Wnt/β-catenin signalling to be unaffected by canonical NF-κB signalling. The observed crosstalk of the simultaneous TNF stimulation on the β-catenin dynamics is unexpected considering the hypothesis that assumes crosstalk through competition of β-catenin and IκB for a limited pool of available β-TrCP.

To explore whether the results observed for the reference parameter set are characteristic for all 400 selected parameter sets, the crosstalk impact (CI) of simultaneous TNF stimulation on β-catenin dynamics and simultaneous Wnt stimulation on NF-κB dynamics are calculated for each parameter set (Equation 2.5, Section 2.5). The CI characterises the maximal difference between two dynamics at a certain time point t in the simulations. A CI ≥ 0.05 is considered to yield an experimentally observable difference (Section 2.5). The analysis shows that none of the parameter sets yields a CI ≥ 0.05 for the dynamics of NF-κB comparing single TNF stimulation with simultaneous stimulation with Wnt and TNF (Table 5.3), indicating that Figure 5.7B represents no isolated case. In the case of β-catenin dynamics, 60 parameter sets (i.e. 15% of all parameter sets) yield a CI ≥ 0.05 for the dynamics of β-catenin comparing single Wnt stimulation with simultaneous stimulation with Wnt and TNF (Table 5.3). For these 60 parameter sets, the simultaneous TNF and Wnt stimulation leads to an increase of the signal amplitude of the β-catenin dynamics up to 1.7-fold, the signalling time by up to 2.3-fold, and signal duration by up to 6.4-fold compared to the single Wnt stimulation.

Table 5.3: Fraction of potentially observable crosstalk in the 400 selected parameter sets.

Fraction of parameter sets in all 400 selected parameter sets realising crosstalk that may be observable in the β-catenin or the NF-κB dynamics. Crosstalk is considered to be observable if the calculated crosstalk impact (CI) is greater than or equal to 0.05 (Section 2.5).

	β-catenin dynamics		NF-κB dynamics	
	CI < 0.05	CI ≥ 0.05	CI < 0.05	CI ≥ 0.05
Count	340	60	400	0
Fraction	85%	15%	100%	0%

5.3.5 Influence of parameter choice on crosstalk of TNF to Wnt/β-catenin signalling

The analysis in Section 5.3.4 has shown that the choice of the parameter set influences the CI. To investigate to what extend the parameter values influence the crosstalk from TNF to Wnt/β-catenin signalling, Kendall rank correlation coefficients for all 400 selected parameter sets are calculated. The steady state equations of β-catenin and NF-κB (Equation 9.17 and Equation 9.23, respectively, in Appendix 9.1) show that several model parameters are linked to each other at steady state conditions. That is, certain parameters appear in association with always the same other parameters in Equation 9.17 and Equation 9.23. Thus, they can be grouped into parameter combinations ($\frac{k_{18}}{k_{16}}$, $\frac{k_{18}}{k_{13}}$, $\frac{k_{17}}{k_{17r}+k_{18}}$, $\frac{k_{12} \cdot K_{15}}{vmax_{11}}$, $\frac{vmax_{15}}{k_{16}}$, $\frac{k_{13r}}{k_{13}}$, $\frac{v_9}{k_{10}}$, $\frac{k_8}{k_3}$, $\frac{k_7}{k_{7r}+k_8}$, $\frac{k_3}{k_6}$, and $\frac{k_{3r}}{k_3}$). The following Kendall rank correlation analysis considers these parameter combinations in addition to the individual parameters. Note that the six individual parameters that are fixed in all 400 selected parameter sets (v_1, k_2, v_5, k_6, k_{12}, and k_{16}) are not taken into account because rank correlations have no meaning in these cases. The analysis reveals that the parameter v_9 and the parameter combination $\frac{k_{17}}{k_{17r}+k_{18}}$ most strongly correlate with the values of CI. Their respective Kendall rank correlation coefficients are -0.53 and 0.48 (Table 9.3, Appendix 9.4). The positive Kendall rank correlation coefficient of the parameter combination $\frac{k_{17}}{k_{17r}+k_{18}}$ of 0.48 indicates that the more the formation of NF-κB/IκB/β-TrCP complexes (Reaction 17, Figure 5.1) dominates the dissociation of NF-κB/IκB/β-TrCP complexes (reverse Reaction 17 and Reaction 18, Figure 5.1) the larger is the CI of TNF stimulation on the β-catenin response dynamics upon Wnt stimulation. A possible molecular mechanistic interpretation is that the larger the value of the parameter combination $\frac{k_{17}}{k_{17r}+k_{18}}$, the more β-TrCP is sequestered into NF-κB/IκB/β-TrCP complexes. Consequently, less β-TrCP is available for the Wnt/β-catenin signalling pathway resulting in an observable crosstalk of simultaneous TNF stimulation on β-catenin dynamics. A further analysis shows however that large values of the parameter combination $\frac{k_{17}}{k_{17r}+k_{18}}$ are not sufficient to observe such crosstalk (Figure 5.9A). In Figure 5.9A, the values of the parameter combination $\frac{k_{17}}{k_{17r}+k_{18}}$ and the corresponding CI are plotted for all 400 selected parameter sets. It shows that values of $\frac{k_{17}}{k_{17r}+k_{18}} > 10^{-3}$ nM^{-1} are associated with CI values above as well as below 0.05 (indicated by grey line in Figure 5.9A).

A possible explanation why values of the parameter combination $\frac{k_{17}}{k_{17r}+k_{18}} > 10^{-3}$ nM^{-1} are not sufficient to observe crosstalk could be that fast β-TrCP production (Reaction 9 in Figure 5.1) might balance the changes in β-TrCP abundance due to its sequestration into NF-κB/IκB/β-TrCP complexes.

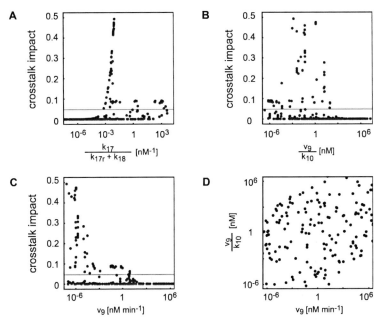

Figure 5.9: Influence of parameters and parameter combinations on the crosstalk impact of simultaneous TNF stimulation on β-catenin dynamics.

(A - C) Crosstalk impact of simultaneous TNF stimulation on β-catenin dynamics is calculated for all 400 selected parameter sets and plotted versus the respective values of (A) $\frac{k_{17}}{k_{17r}+k_{18}}$, (B) $\frac{v_9}{k_{10}}$, and (C) v_9. Grey lines indicate the threshold of observable crosstalk impacts. Crosstalk is considered to be observable if the calculated crosstalk impact is greater than or equal to 0.05 (Section 2.5). (D) β-TrCP steady state ($\frac{v_9}{k_{10}}$) versus production rate constant of β-TrCP (v_9) are plotted for all 400 selected parameter sets.

Indeed, the parameter of β-TrCP production (v_9) has the largest Kendall rank correlation coefficients in absolute values (Table 9.3, Appendix 9.4). The negative sign of the Kendall rank correlation coefficient of parameter v_9 indicates that small values of v_9 (i.e., slow β-TrCP production) correlate with larger CI. Further analysis shows that small values of v_9 ($v_9 <$ 10 nM·min^{-1}) are also not sufficient to observe crosstalk since they yield CI values above and below 0.05 (Figure 5.9C).

Low values of parameter v_9 do not only yield slow rates of β-TrCP production but may also influence the steady state concentration of β-TrCP, since the steady state is given by the ratio of v_9 to k_{10} (Equation 9.11, Appendix 9.1). To explore, how strong the production rate constant of β-TrCP (v_9) correlates with the steady state concentration of β-TrCP ($\frac{v_9}{k_{10}}$), their values are plottet for all 400 selected parameter sets (Figure 5.9D). The analysis shows a small correlation between v_9 and steady state concentration of β-TrCP having a Kendall rank correlation coefficient of 0.25. Therefore in a next step, the rank correlation of the steady state concentration of β-TrCP and the CI is investigated. The Kendall rank correlation coefficient of $\frac{v_9}{k_{10}}$ is -0.35, which is about two thirds of that of individual parameter v_9 (Table 9.3, Appendix 9.4). The values of $\frac{v_9}{k_{10}}$ and corresponding CI for all 400 selected parameter sets are plotted in Figure 5.9B. The figure shows that steady state concentrations of β-TrCP smaller than approximately 10^2 nM are necessary but not sufficient for crosstalk of simultaneous TNF stimulation on the β-catenin dynamics upon Wnt stimulation. If the steady state concentration is larger than 10^2 nM, no CI greater than or equal to 0.05 are found (Figure 5.9B). In these cases, probably the competition between the DC/β-catenin and NF-κB/IκB complexes for β-TrCP binding has ceased.

Taken together, the Kendall rank correlation analysis indicates that the production rate of β-TrCP (parameter v_9) and the parameter combination $\frac{k_{17}}{k_{17r}+k_{18}}$ most strongly correlate with the CI of TNF stimulation on β-catenin dynamics.

5.3.6 Effects of the reduction of production and degradation rates of β-TrCP on the impact of TNF stimulation on Wnt/β-catenin signalling

In Section 5.3.5, it was shown that the β-TrCP production rate constant v_9 correlates the strongest with the crosstalk impact of TNF stimulation on β-catenin dynamics. It might be speculated that a deceleration of the β-TrCP production could enhance the observable crosstalk because the deficit of unbound β-TrCP due to its sequestration is not readily compensated by β-TrCP synthesis. To test this hypothesis, the respective CI is calculated for a reduced production rate constant of β-TrCP (v_9) is in all 400 selected parameter sets. To obtain a new reduced parameter of β-TrCP production, v_9^{new} is calculated using Equation 5.6:

$$v_9^{new} = \frac{1}{\tau} \cdot \frac{(\beta - TrCP)_{stst}}{1 + (\beta - TrCP)_{stst}} \qquad\qquad 5.6$$

Using Equation 5.6, a maximal possible new β-TrCP production rate constant ($\frac{1}{\tau}$) can be chosen, which is always the upper limit for any β-TrCP steady state *(β-TrCP)*$_{stst}$ given by the 400 selected parameter sets. In order to reduce the β-TrCP production rate constant without changing the steady state concentration of β-TrCP, also a new β-TrCP degradation rate parameter k_{10}^{new} has to be calculated (Equation 5.7).

$$k_{10}^{new} = \frac{1}{\tau} \cdot \frac{1}{1 + (\beta - TrCP)_{stst}} \qquad\qquad 5.7$$

Here again, the parameter ratio $\frac{1}{\tau}$ sets the upper limit of the new rate constant for any given β-TrCP steady state (*(β-TrCP)*$_{stst}$). The ratio of parameter v_9^{new} and parameter k_{10}^{new} gives the β-TrCP steady state concentration and is thus identical to the ratio of the original parameters v_9 and k_{10} (Equation 9.11, Appendix 9.1). Note that τ has some analogy to the time constant of a metabolic interconversion reaction with linear rate laws (Heinrich and Schuster, 1996). Substituting *(β-TrCP)*$_{stst}$ in Equation 5.7 by $\frac{v_9}{k_{10}}$ and solving for τ yields Equation 5.8.

$$\tau = \frac{1}{v_9 + k_{10}} \qquad\qquad 5.8$$

Thus, τ may be considered to characterise the timescale of the dynamics of β-TrCP production and degradation. In order to obtain dynamics of β-TrCP production and degradation that are slower than the dynamics of β-catenin and NF-κB, τ is fixed at 10^3 min. This way, τ is one to two orders of magnitude larger than the signalling time of β-catenin and NF-κB upon simultaneous Wnt and TNF stimulation for the reference parameter set (286 min and 75 min, respectively).

The parameters v_9^{new} and k_{10}^{new} are calculated for all 400 selected parameter sets using Equation 5.6 and Equation 5.7, respectively. The corresponding new CIs are summarised in Table 5.4. A comparison of the fraction of $CI \geq 0.05$ in the case of β-catenin dynamics in Table 5.4 with that in Table 5.3, where the original parameters v_9 and k_{10} were used, reveals a 2.25-fold increase of the fraction of parameter sets that allow for a $CI \geq 0.05$ if the production and degradation of β-TrCP is slowed down. The parameter sets of Table 5.3 that yield $CI \geq 0.05$ also yield $CI \geq 0.05$ for the reduced parameters v_9^{new} and k_{10}^{new}. In the case of NF-κB dynamics, the reduction of β-TrCP production and degradation led to a detection of $CI \geq 0.05$ for five parameter sets (Table 5.4). In the limit condition that production and degradation of β-TrCP are more than one order of magnitude slower than the signalling time of the dynamics of β-catenin and NF-κB, a conservation relation of total β-TrCP can be assumed. This is possible, because under this condition the production and degradation of β-TrCP hardly influence the concentration of unbound β-TrCP in the time frame of β-catenin and NF-κB dynamics. In the case of total β-TrCP conservation, sequestration of β-TrCP into NF-κB/IκB/β-TrCP complexes due to TNF stimulation reduces the β-TrCP concentration available for the Wnt/β-catenin pathway, which may affect the concentration of the DC/β-catenin/β-TrCP complexes. Similarly, Wnt stimulation may influence the dynamics of NF-κB/IκB/β-TrCP complexes by changing the distribution of total β-TrCP among the DC/β-catenin/β-TrCP and NF-κB/IκB/β-TrCP complexes and unbound β-TrCP.

Table 5.4: Fraction of observable crosstalk in the 400 selected parameter sets with reduced production and degradation rates of β-TrCP.

Fraction of crosstalk in all 400 parameter sets that may be observable in the β-catenin or the NF-κB dynamics. The reduced β-TrCP production and degradation rate constants (v_9^{new} and k_{10}^{new}, respectively) were consider in all 400 parameter sets for the simulations. Crosstalk is considered to be observable if the calculated crosstalk impact (CI) is greater than or equal to 0.05 (Section 2.5).

	β-catenin dynamics		NF-κB dynamics	
	CI < 0.05	CI ≥ 0.05	CI < 0.05	CI ≥ 0.05
Count	265	135	395	5
Fraction	66%	34%	99%	1%

Despite the 2.25-fold increase in the fraction of parameter sets that yield a CI ≥ 0.05 in case of β-catenin dynamics, the majority of about 66% of the 400 parameter sets (Table 5.4) still do not reveal observable crosstalk. To investigate the underlying reason, the Kendall rank correlation coefficients between the individual parameters as well as the parameter combinations and the values of CI are calculated for the 400 parameter sets with reduced production (v_9^{new}) and degradation of β-TrCP (k_{10}^{new}) (Table 9.4, Appendix 9.4). The analysis reveals that the parameter combination $\frac{k_{17}}{k_{17r}+k_{18}}$ most strongly correlates with the value of CI. The Kendall rank correlation coefficient is 0.96 (Table 9.4, Appendix 9.4). For further analyses, the parameter combination $\frac{k_{17}}{k_{17r}+k_{18}}$ is plotted with the corresponding value of CI for all 400 parameter sets with reduced production and degradation of β-TrCP (Figure 5.10A). Figure 5.10A indicates that a critical value of the parameter combination $\frac{k_{17}}{k_{17r}+k_{18}}$ of approximately $1.1 \cdot 10^{-3}$ nM^{-1} exists to detect an observable crosstalk impact CI ≥ 0.05.

For the analysis in this section, the reduced β-TrCP production and degradation rate constants v_9^{new} and k_{10}^{new} are calculated for all 400 selected parameter sets using Equation 5.6 and Equation 5.7, respectively. This reduction of the original parameters v_9 and k_{10} is done without changing their ratio, that is, without changing the steady state concentration of β-TrCP. Consequently, the findings of Section 5.3.5 considering the correlation of the abundance of β-TrCP ($\frac{v_9}{k_{10}}$) and the CI for the original 400 selected parameter sets can be retrieved in analysis of the rescaled 400 parameter sets as well. In particular, the abundance of β-TrCP ($\frac{v_9^{new}}{k_{10}^{new}}$) has an upper limit to detect observable crosstalk of about 10^2 nM (Figure 5.10B). This indicates that in the case of β-TrCP abundance larger than 10^2 nM probably no competition for β-TrCP between the canonical NF-κB and Wnt/β-catenin signalling pathway takes place. The analysis furthermore shows that β-TrCP abundance below this limit is still not a sufficient condition to detect observable crosstalk (CI ≥ 0.05). For steady state concentrations smaller than 10^2 nM, CI values above as well as below 0.05 (indicated by grey line in Figure 5.10B) are found.

Taken together, the 2.25-fold increase in the fraction of parameter sets that yield a CI ≥ 0.05 in the β-catenin dynamics is achieved solely by reducing the production and degradation rate constants of β-TrCP (v_9 and k_{10}, respectively) without changing the steady state concentration of β-TrCP. The analyses suggest that in order to realise observable crosstalk, the parameter $\frac{k_{17}}{k_{17r}+k_{18}}$ has to pass a critical minimal value of 10^{-3} nM^{-1} and the production and degradation rate of β-TrCP needs to be sufficiently slow ($\frac{1}{\tau}$ less than 10^{-3} min^{-1}, see Equation 5.8). The steady state of β-TrCP also needs to be lower than a critical threshold of 10^{2} nM to allow for competition for β-TrCP. However, a low abundance of β-TrCP is not sufficient to observe crosstalk.

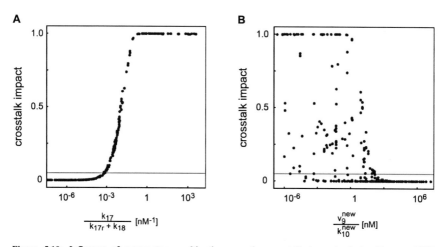

Figure 5.10: Influence of parameter combinations on the crosstalk impact of simultaneous TNF stimulation on β-catenin dynamics with reduced production and degradation rates of β-TrCP.
Reduced production and degradation rates of β-TrCP in combination with a value of $\frac{k_{17}}{k_{17r}+k_{18}}$ larger than 10^{-3} nM^{-1}, but not with low β-TrCP abundance ($\frac{v_9^{new}}{k_{10}^{new}} < 10^{2}$ nM), is sufficient to observe crosstalk of TNF stimulation on β-catenin dynamics. Crosstalk impact of simultaneous TNF stimulation on β-catenin dynamics is plotted versus (A) the respective value of $\frac{k_{17}}{k_{17r}+k_{18}}$, and (B) the respective value of $\frac{v_9^{new}}{k_{10}^{new}}$. Simulations were performed for all 400 parameter sets with reduced production and degradation rate constants of β-TrCP (v_9^{new} and k_{10}^{new}, respectively). Note that the variation in β-TrCP steady state ($\frac{v_9^{new}}{k_{10}^{new}}$) in (B) is due to variation in the 400 parameter sets (see Figure 5.5A) and not caused by the reduction of the original parameters v_9 and k_{10} to v_9^{new} using Equation 5.6 and k_{10}^{new} using Equation 5.7, respectively. Grey lines indicate the threshold of observable crosstalk impacts. Crosstalk is considered to be observable if the calculated crosstalk impact is greater than or equal to 0.05 (Section 2.5).

5.4   Discussion

In this chapter, parametric conditions that realise or prevent potential crosstalk between Wnt/β-catenin and canonical NF-κB signalling by competitive β-TrCP sequestration were explored. To this end, a minimal model of competitive β-TrCP sequestration was developed (Figure 5.1). It explicitly incorporates the dynamics of β-TrCP. β-TrCP may be sequestered into either the module of Wnt/β-catenin signalling or that of canonical NF-κB signalling. Model simulations revealed that fast sequestration of β-TrCP into the NF-κB/IκB/β-TrCP complex (large value of parameter k_{17}) and retention of β-TrCP in this complex (small value of the sum of the parameters k_{17r} and k_{18}) are important conditions to realise crosstalk of simultaneous TNF stimulation on β-catenin dynamics (Section 5.3.5). However, the analysis showed that a large value of the parameter combination $\frac{k_{17}}{k_{17r}+k_{18}}$ on its own is not sufficient to enable observable crosstalk (Figure 5.9A). Values of the parameter combination $\frac{k_{17}}{k_{17r}+k_{18}} > 10^{-3}$ nM^{-1} were necessary to detect observable crosstalk (CI ≥ 0.05), however also CI values less than 0.05 were calculated indicating no observable crosstalk. A possible explanation may be that the TNF-stimulus-induced reduction of the β-TrCP fraction, which remains available for Wnt/β-catenin signalling, may be compensated by a fast production of β-TrCP (large β-TrCP production rate v_9). Indeed, model simulations in Section 5.3.6 showed that a reduction of production and degradation rate of β-TrCP strongly influences the observability of potential crosstalk (Table 5.4).

In general, the existence of a β-TrCP production and degradation reaction in the model (Reaction 9 and Reaction 10 in Figure 5.1, respectively) separated the model parameters into three distinct groups with respect to their influence on the model species at steady state condition. The first group included all parameters of the Wnt/β-catenin signalling module (red in Figure 5.1). They only influenced the species of the Wnt/β-catenin signalling module but not those of the canonical NF-κB signalling module (blue in Figure 5.1). The second group contained all parameters of the canonical NF-κB signalling module. They only influenced the species of the canonical NF-κB signalling module but not those of the Wnt/β-catenin signalling module. In consequence, a Wnt stimulus did not affect the steady state concentration of NF-κB nor did a TNF stimulus influence the steady state concentration of β-catenin. This result agrees with similar observations in other minimal models of crosstalk via shared components (Seaton and Krishnan, 2011). The third group of parameters included the β-TrCP production rate (v_9) and degradation rate (k_{10}), which affected the steady state of β-TrCP as well as the canonical NF-κB signalling module and the Wnt/β-catenin signalling module. Changes of β-TrCP abundance by modulation of its production (v_9) and/or degradation rate (k_{10}) influenced the steady states of NF-κB and β-catenin in opposite directions (Figure 5.5B). The opposite influence is caused by the different mechanistic dependences of β-catenin and NF-κB on β-TrCP. β-TrCP directly regulates β-catenin degradation and hence negatively affects the steady state concentration of β-catenin. In contrast, β-TrCP mediates the degradation of the inhibitor of NF-κB (IκB) and thus has a positive (double negative) influence on the steady state concentration of NF-κB. The opposite regulation observed in the model simulations is in agreement with experimental reports. For instance, overexpression experiments showed that up-regulation of β-TrCP in various mammalian cell types down-regulated β-catenin and enhanced the activity of NF-κB (Bhatia et al., 2002; Fuchs et al., 1999; Hart et al., 1999; Kroll et al., 1999; Muerkoster et al., 2005; Spiegelman et al., 2000; Spiegelman et al., 2001; Wang et al., 2004).

As mentioned above, a Wnt stimulus did not affect the steady state concentration of NF-κB nor did a TNF stimulus influence the steady state concentration of β-catenin. Despite that, a Wnt stimulus may influence the transient dynamics of NF-κB upon TNF stimulation and a TNF stimulus may impact the transient dynamics of β-catenin upon Wnt. To investigate this hypothesis, simulations of NF-κB and β-catenin upon simultaneous stimulation with TNF and Wnt were compared to dynamics upon single TNF or Wnt stimulation. The analysis of the minimal model demonstrated that the crosstalk impact on the dynamics is influenced by the production and degradation rate of β-TrCP (v_9 and k_{10}) provided a sufficiently large value of the parameter combination $\frac{k_{17}}{k_{17r}+k_{18}} > 10^{-3}$ nM^{-1} (β-TrCP sequestration to and release from the NF-κB/IκB complex) and an abundance of β-TrCP of less than 10^2 nM (Figure 5.9).

In the limit of production and degradation of β-TrCP having much slower dynamics than the dynamics of Wnt/β-catenin and canonical NF-κB signal transduction, a conservation relation of total β-TrCP can be assumed. In this limit, the analysis revealed that values of $\frac{k_{17}}{k_{17r}+k_{18}}$ greater than the critical value of 10^{-3} nM^{-1} allow for observable crosstalk of simultaneous TNF stimulation on Wnt/β-catenin signalling (Figure 5.10A). For values of $\frac{k_{17}}{k_{17r}+k_{18}}$ above the critical value, reduced production and degradation rates of β-TrCP generally increase the impact of simultaneous TNF stimulation on β-catenin dynamics upon Wnt stimulation. In contrast, if production and degradation of β-TrCP is much faster than the dynamics of Wnt/β-catenin and canonical NF-κB signal transduction, changes in β-TrCP abundance due to sequestration effects are instantly compensated by fast resynthesis of sequestered β-TrCP or fast degradation of dissociated β-TrCP. Hence, fast production and degradation rates of β-TrCP render the Wnt/β-catenin and canonical NF-κB signalling modules independent of each other and it is unlikely to observe crosstalk in this scenario.

β-TrCP is reported to be an unstable protein that has a fast turnover, i.e. fast production and degradation (Fuchs et al., 1999). It has been suggested by experiments that β-TrCP dimerises and mediates its own degradation by auto-ubiquitination (Deshaies and Joazeiro, 2009; Suzuki et al., 2000). The analysis of the minimal model predicts that fast production (Figure 5.9C) and fast degradation of β-TrCP (not shown) prevents crosstalk via competitive β-TrCP sequestration. Thus, unstable β-TrCP seems to contradict the observations of crosstalk in Vpu overexpression experiments (Besnard-Guerin et al., 2004; Bour et al., 2001). The simulations of the minimal model further indicated that deceleration of the production and degradation of β-TrCP may enable crosstalk (Table 5.4). Indeed, the degradation of several F-box proteins is attenuated by the binding of their respective target proteins (de Bie and Ciechanover, 2011; Li et al., 2004). In the case of HOS, increasing concentrations of phosphorylated IκB were suggested to protect HOS from degradation (Fuchs et al., 2004). It might be speculated that the sequestration of β-TrCP by its target proteins reduces its fast autocatalytic degradation, and thereby provides conditions that potentially allow for crosstalk. In this respect, it is interesting to note that particularly overexpression of Vpu or the transcription factor ATF (both are target proteins of β-TrCP) led to the accumulation of β-catenin due to competitive β-TrCP sequestration (Besnard-Guerin et al., 2004; Bour et al., 2001).

The analysis of the minimal model revealed an upper limit of 10^2 nM for the steady state of β-TrCP that allows for potential crosstalk of simultaneous TNF stimulation to transient response dynamics of β-catenin upon Wnt stimulation (Figure 5.10B). It was expected to detect an upper limit considering the working hypothesis that crosstalk arises from competition of β-TrCP targets for a limited pool of β-TrCP. It also agrees with the experimental observation that overexpression of β-TrCP inhibits the influence of overexpressed Vpu on β-catenin (Besnard-Guerin et al., 2004). However, the limit concentration of β-TrCP is approximately 100 nM in the model (Figure 5.10B), which is larger than the steady state concentrations of all other species of the model. A limited pool of β-TrCP, i.e. very low abundance of β-TrCP relative to other signalling pathway components, seems therefore not necessary to enable crosstalk via competitive β-TrCP sequestration. This finding is further supported by the observation that crosstalk is still possible in the simulations despite a presence of more than 60% of the total β-TrCP freely available for sequestration (Figure 5.8). Indeed, HOS knock-down experiments in FWD1 knock-out mice have shown that 20% residual HOS are sufficient to generally maintain normal tissue homeostasis (except for testis) (Kanarek et al., 2010). The model analysis furthermore showed that low abundance of β-TrCP is not sufficient to observe crosstalk (Figure 5.10B). In contrast, a value of the parameter combination $\frac{k_{17}}{k_{17r}+k_{18}}$ larger than 10^{-3} nM^{-1} in combination with slow β-TrCP production and degradation (less than one order of magnitude than the signalling time of β-catenin dynamics) is sufficient to enable observable crosstalk (Figure 5.10A).

Taken together, the minimal model reproduced the dynamics of central species of the detailed model of Wnt/β-catenin signalling in response to Wnt stimulation and those of the detailed model of canonical NF-κB signalling upon TNF stimulation (Figure 5.3). It thus provides a useful tool to investigate the conditions that may allow for crosstalk between the Wnt/β-catenin and canonical NF-κB signalling pathway via competitive β-TrCP sequestration. The analysis of the minimal model revealed that β-TrCP sequestration into and release from the NF-κB/IκB complex (Reaction 17 and Reaction 18 in Figure 5.1), rather than total β-TrCP abundance, strongly correlate with the crosstalk impact of simultaneous TNF stimulation on Wnt-stimulus-induced β-catenin dynamics (Table 9.3, Appendix 9.4), especially in the case of slow β-TrCP production and degradation (Figure 5.10). Of note, observable crosstalk between the Wnt/β-catenin and canonical NF-κB signalling pathway via competitive β-TrCP sequestration seemed to be unidirectional from TNF to Wnt/β-catenin signalling in the minimal model (Table 5.3). This finding contradicts with experimental reports that show β-catenin-induced activation of NF-κB in human embryonic kidney 293T cells and mouse embryonic fibroblast NIH3T3 cells (Fuchs et al., 2004; Spiegelman et al., 2000; Spiegelman et al., 2002b). To explore the mechanisms influencing the direction of crosstalk in the minimal model in further detail, two incipient investigations have been initiated, which are briefly discussed in Section 5.5. In particular, the consequences of changes in the relative timing between the two pathways (Section 5.5.2) and in the absolute concentrations of their respective pathway components (Section 5.5.3) on the direction of the crosstalk are analysed.

5.5 Preliminary results: conditions that influence the direction of crosstalk between canonical NF-κB and Wnt/β-catenin signalling

5.5.1 Introduction

In the last decade, several theoretical and experimental studies have been undertaken to explore cellular strategies that ensure specificity in the transduction of a given input signal to a desired output through complex signalling pathway networks (Furukawa and Hohmann, 2013; Schwartz and Madhani, 2004; Thalhauser and Komarova, 2009). A great portion of complexity in these networks arises from signalling components that are shared by multiple signalling pathways; for instance, β-TrCP in the present study. Sharing of signalling components between pathways offers a mechanistic mean to integrate multiple environmental signals and coordinate corresponding signalling pathway responses (McNeill and Woodgett, 2010). But it also demands for regulatory mechanisms to prevent undesired cross-activation between pathways by a transduced signal, hereafter called insulation mechanisms. A great advance in the understanding of crosstalk and insulation mechanisms has been gained in theoretical and experimental studies of MAPK signalling pathways in yeast and PC12 rat-derived neural progenitor cells (Furukawa and Hohmann, 2013; Saito, 2010; Schwartz and Madhani, 2004; Thalhauser and Komarova, 2009). In theoretical studies, several insulation mechanisms have been proposed: combinatorial signalling, cross-pathway inhibition, scaffolding or compartmentation of the shared signalling component, temporal separation, and kinetic insulation (Bardwell et al., 2007; Behar et al., 2007; Haney et al., 2010; Hu et al., 2011; Saito, 2010; Schaber et al., 2006; Schwartz and Madhani, 2004).

In the case of the minimal model, the canonical NF-κB signalling pathway seemed protected from crosstalk of Wnt/β-catenin signalling via competitive β-TrCP sequestration (Table 5.3). To find possible mechanistic explanations for this observation, two initial working hypotheses were formulated: (i) the unidirectional crosstalk may be due to kinetic insulation of both pathways, and (ii) insulation may be achieved through differences in the abundance of pathway components. Thus, the investigations presented in Section 5.5.2 focus on the consequences of changes in the relative timing between the canonical NF-κB and Wnt/β-catenin signalling module on their crosstalk. In Section 5.5.3, the consequences of changes in the absolute concentrations of the pathway components of the canonical NF-κB and Wnt/β-catenin signalling module on the direction of the crosstalk are investigated. Note that the analyses are restricted to the reference parameter set (Table 7.10 and Table 7.11, Appendix 7.4).

5.5.2 Effects of variations in signalling time on the direction of crosstalk

The Wnt/β-catenin and canonical NF-κB pathway transduce their signals on very different time scales. To demonstrate this difference, the signalling time (Equation 2.1, Section 2.4) of β-catenin and NF-κB dynamics is calculated for the reference parameter set (Table 7.10 and Table 7.11, Appendix 7.4). Indeed, the signalling time of the β-catenin dynamics upon Wnt stimulation (285.9 min; Table 5.5, Figure 5.11A) is much larger than that of the NF-κB dynamics upon TNF stimulation (75.0 min; Table 5.5, Figure 5.11D). In consequence, Wnt-stimulus-dependent changes in β-TrCP concentration may occur to slow for canonical NF-κB signalling to affect NF-κB dynamics. To test this hypothesis, the dynamics of each pathway module was rescaled in time (see Appendix 9.2). Three different cases of relative timing between the two signalling modules are considered: (i) the signalling time of β-catenin

Table 5.5: Quantification of dynamical measures of β-catenin and NF-κB dynamics on different relative time scales.

Quantification of signalling time, signal amplitude, and signal duration (Section 2.4) of β-catenin and NF-κB dynamics on different relative time scales shown in Figure 5.11. The measures correspond to the dynamics of β-catenin (Figure 5.11A – C) and NF-κB (Figure 5.11D – F) in response to single Wnt or TNF stimulation, respectively. The numbers in brackets state the fold-change between the dynamics of simultaneous Wnt and TNF stimulation and corresponding pathway-specific stimulation.

	Figure 5.11A	Figure 5.11B	Figure 5.11C
Signalling time of β-catenin	285.9 min (1.89)	84.1 min (3.91)	75.0 min (3.64)
Signal amplitude of β-catenin	22.5 nM (1.68)	22.5 nM (1.82)	22.5 nM (2.43)
Signal duration of β-catenin	228.8 min (5.89)	67.3 min (18.9)	60.1 min (17.7)
	Figure 5.11D	Figure 5.11E	Figure 5.11F
Signalling time of NF-κB	75.0 min (1.00)	85.3 min (1.00)	285.5 min (1.00)
Signal amplitude of NF-κB	54.5 nM (1.00)	54.5 nM (1.00)	54.5 nM (1.00)
Signal duration of NF-κB	469.0 min (1.00)	469.6 min (1.00)	537.8 min (1.00)

dynamics is larger than that of NF-κB dynamics (reference situation, Figure 5.11A and D, respectively), (ii) the signalling time of β-catenin and NF-κB dynamics are similar (Figure 5.11B and E, respectively), and (iii) the signalling time of β-catenin dynamics is smaller than that of NF-κB dynamics (Figure 5.11C and F, respectively). The quantification of signal amplitude, signalling time, and signal duration (Section 2.4) of β-catenin and NF-κB dynamics upon pathway-specific stimulation are summarised in Table 5.5 for the three cases (i), (ii), and (iii). The numbers in brackets state the fold-change between the dynamics of simultaneous Wnt and TNF stimulation and corresponding pathway-specific stimulation.

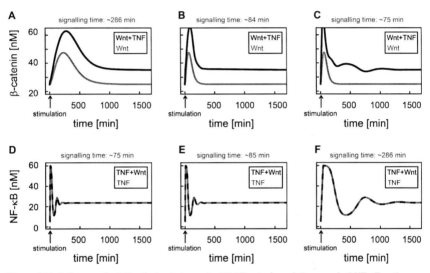

Figure 5.11: Influence of relative timing between the Wnt/β-catenin and the canonical NF-κB pathway on their crosstalk impact.

(A - C) Dynamics of β-catenin upon single Wnt stimulation (grey lines) or simultaneous Wnt and TNF stimulation (black lines) are compared. (D - F) Dynamics of NF-κB upon single TNF stimulation (grey lines) or simultaneous TNF and Wnt stimulation (black dashed lines) are compared. Note that both NF-κB dynamics lay on top of each other. The time scales of the Wnt/β-catenin signalling module and the canonical NF-κB signalling module are rescaled (see Appendix 9.2) to realise three different cases of relative timing: (A, D) the signalling time of β-catenin dynamics is larger than that of NF-κB dynamics (reference situation), (B, E) the signalling time of β-catenin and NF-κB dynamics are similar, and (C, F) the signalling time of β-catenin dynamics is smaller than that of NF-κB dynamics. The respective signalling times are provided above each panel.

In general, the rescaling in time influences signalling time and signal duration but does not change the signal amplitude of the β-catenin (Table 5.5, Figure 5.11A – C) and NF-κB dynamics (Table 5.5, Figure 5.11D – F) upon pathway-specific stimulation (grey lines in Figure 5.11). In the case of simultaneous TNF and Wnt stimulation (black lines in Figure 5.11), signalling time, signal duration, and signal amplitude of the β-catenin dynamics are yet again varied compared to single Wnt stimulation. This is indicated by the fold-change (values in brackets in Table 5.5, Figure 5.11A – C) of signalling time, signal duration, and signal amplitude, that is always larger than 1. In contrast, the NF-κB dynamics are unaffected by simultaneous TNF and Wnt stimulation indicated by a fold-change of 1 in case of signalling time, signal duration, and signal amplitude (values in brackets in Table 5.5, Figure 5.11D – F). The latter result demonstrates that changes in the relative timing of the two pathway modules do not result in a detectable crosstalk impact of simultaneous Wnt and TNF stimulation on NF-κB dynamics. In contrast, the β-catenin dynamics show crosstalk upon simultaneous TNF and Wnt stimulation in all three scenarios of relative timing (Figure 5.11A – C). With decreasing signalling times of β-catenin dynamics and increasing signalling times of NF-κB dynamics, the larger is the fold-change of signal amplitude (1.68 to 2.43-fold; Table 5.5, Figure 5.11A – C). Similarly, the fold-change of signalling time and signal duration of β-catenin changes between all three scenarios of relative timing (Table 5.5, Figure 5.11A – C) although a trend as for signal amplitude cannot be identified.

Taken together, the rescaling in the relative timing affects signal amplitude, signalling time, and signal duration of the β-catenin dynamics for the reference parameter set (Table 7.10 and Table 7.11, Appendix 7.4), but does not establish crosstalk of Wnt stimulation to NF-κB dynamics.

5.5.3 Effects of scaling the relative concentrations of the pathway components on the direction of crosstalk

To further explore possible reasons for absence of observable crosstalk of simultaneous Wnt stimulation on NF-κB dynamics, the dynamics of β-TrCP, the NF-κB/IκB/β-TrCP complex, and the DC/β-catenin/β-TrCP complex upon simultaneous Wnt and TNF stimulation are investigated (Figure 5.12). The simulations reveal for the reference parameter set that the concentration of the DC/β-catenin/β-TrCP complex (Figure 5.12A) is much smaller than the concentrations of β-TrCP and the NF-κB/IκB/β-TrCP complex (Figure 5.12B and C, respectively). It might be that Wnt-stimulus-dependent changes in the concentration of the DC/β-catenin/β-TrCP complex are too small to influence the dynamics of β-TrCP and the NF-κB/IκB/β-TrCP complex. Crosstalk of simultaneous Wnt stimulation to the canonical NF-κB signalling is therefore impossible in the case of the reference parameter set. A re-examination of the five parameter samples that yield a CI of the NF-κB dynamics greater than 0.05 (Table 5.4) shows that the ratio of the absolute stimulated steady state concentrations of the DC/β-catenin/β-TrCP complex and the NF-κB/IκB/β-TrCP complex upon simultaneous Wnt and TNF stimulation is close to one in all five cases. These observations suggest that the relative concentration the DC/β-catenin/β-TrCP complex and the NF-κB/IκB/β-TrCP complex may influence the direction of potential crosstalk.

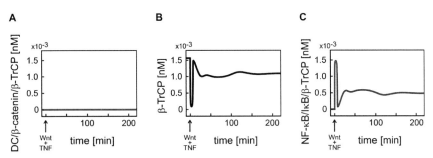

Figure 5.12: Dynamics of β-TrCP and its complexes in the Wnt/β-catenin and canonical NF-κB pathway module upon simultaneous Wnt and TNF stimulation.

Dynamics of (A) the DC/β-catenin/β-TrCP complex, (B) β-TrCP, and (C) the NF-κB/IκB/β-TrCP complex upon simultaneous Wnt and TNF stimulation are simulated using the reference parameter set (Table 7.10 and Table 7.11, Appendix 7.4).

To support this hypothesis in a proof of principle approach, the reference parameter set is rescaled (see Appendix 9.3) in order to obtain similar stimulated steady state concentrations of the NF-κB/IκB/β-TrCP complex and the DC/β-catenin/β-TrCP complex ($1.3 \cdot 10^{-9}$ nM and $4.4 \cdot 10^{-8}$ nM, respectively). The analysis of the β-catenin and NF-κB dynamics for this new rescaled parameter set reveals crosstalk in both pathways (Figure 5.13). Simultaneous stimulation with TNF and Wnt increases the signal amplitude, signalling time, and signal duration of the β-catenin dynamics by approximately 1.5-fold, 1.6-fold, and 3.2-fold, respectively. Considering the dynamics of NF-κB for the new rescaled parameter set (Figure 5.13B), simultaneous stimulation with Wnt and TNF results in an observable change in the NF-κB dynamics (CI is approximately 0.26). The simultaneous Wnt and TNF stimulation increases signalling time and signal duration of the NF-κB dynamics by about 1.16-fold and 1.13-fold, respectively. The signal amplitude of the NF-κB dynamics remains unaffected.

In summary, the rescaling in the relative concentrations of the pathway components by manipulating the reference parameter set (Table 7.10 and Table 7.11, Appendix 7.4) can establish crosstalk from TNF to Wnt/β-catenin signalling as well as from Wnt to the canonical NF-κB signalling module.

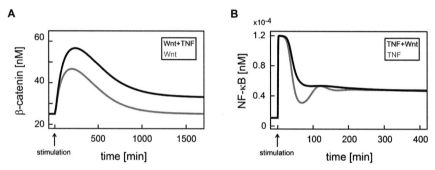

Figure 5.13: Effects of simultaneous TNF and Wnt stimulation on the dynamics of β-catenin and NF-κB, respectively.

(A) Dynamics of β-catenin upon Wnt stimulation (grey line) is compared to dynamics of β-catenin upon simultaneous stimulation with Wnt and TNF (black line). (B) Dynamics of NF-κB upon TNF stimulation (grey dashed line) is compared to dynamics of NF-κB upon simultaneous stimulation with Wnt and TNF (black line). In the simulations the scaled parameter set is considered (Table 7.10 and Table 7.11, Appendix 7.4.4).

5.5.4 Discussion

In the analyses in Section 5.3, a bias in the direction of crosstalk between the Wnt/β-catenin and canonical NF-κB signalling pathway via competitive β-TrCP sequestration was observed (Table 5.3). Generally, NF-κB dynamics upon pathway-specific stimulation and upon simultaneous Wnt and TNF stimulation were superimposable, while β-catenin dynamics responded differently to the two kinds of stimulation in 15% of the parameter sets. In an initial hypothesis, kinetic insulation was assumed to be the reason for the absence of observable crosstalk of Wnt on NF-κB dynamics, because the Wnt/β-catenin and canonical NF-κB pathway transduce their signals on very different time scales. That means that the signalling time of β-catenin dynamics upon Wnt stimulation is much larger than that of the NF-κB dynamics upon TNF stimulation (285.9 min and 75.0 min, respectively; Table 5.5, Figure 5.11A and D). In Section 5.5.2, this hypothesis was disproven although the simulations showed that the relative timing of the pathways strongly influences the crosstalk impact of simultaneous TNF and Wnt stimulation on β-catenin dynamics (Figure 5.11).

Further analysis of the simulations of the DC/β-catenin/β-TrCP and the NF-κB/IκB/β-TrCP complex suggested that their relative concentrations may control the direction of potential crosstalk. To test the new hypothesis, a rescaled parameter set was derived from the reference parameter set (see Appendix 9.3) to adjust the concentrations of the NF-κB/IκB/β-TrCP complex to an about 7.3-fold lower value than that of the concentration of the DC/β-catenin/β-TrCP complex in the minimal model. Using this rescaled parameter set, the simulations showed observable crosstalk of simultaneous Wnt stimulation on NF-κB dynamics in response to TNF stimulation (Figure 5.13B).

This proof-of-concept approach indicates that pathway insulation may also be achieved through differences in the abundances of pathway species. This insulation mechanism was already suggested to provide a functional explanation of the comparatively small concentrations of Axin in the Wnt/β-catenin pathway (Lee et al., 2003). The minimal model may turn out to be useful to closer explore this insulation mechanism.

The analysis emphasised the need for comprehensive quantitative data sets of the species in both pathways in order to verifiably predict the direction of crosstalk in a particular cellular context. It has already been hypothesised that the expression levels of the different target proteins of β-TrCP may influence the competition for β-TrCP (Besnard-Guerin et al., 2004) but has not been addressed by experimental measurements of the involved components yet. Once data becomes available, the minimal model can easily be adapted, rendering it an applicable tool to investigate the competition of the canonical NF-κB and Wnt/β-catenin signalling pathway for β-TrCP.

6. Conclusions and outlook

6.1 Summary

Signalling pathways are regulated through complex often nonlinear interactions between many signalling molecules forming signalling networks. The complex nature of these signalling networks renders it very difficult to understand how a signal propagates through the network and which parameters influence the biologically relevant functional outcome. Mathematical modelling in combination with experimental validation provides a method to gain deeper insights into the mechanisms of signal transduction through the pathways. Usually, an iterative process of experimental data acquisition and testing of theoretical predictions on the one hand, and mathematical modelling and generation of testable predictions on the other hand is performed (Cheong et al., 2008; Klinger et al., 2013; Nguyen et al., 2013; Schaber et al., 2012; Wodke et al., 2013). This thesis constitutes the starting point in such an iterative process by developing and analysing mathematical models of canonical NF-κB and Wnt/β-catenin signalling.

The thesis addressed the regulation of signal transduction through the Wnt/β-catenin and the canonical NF-κB signalling pathway by β-TrCP. β-TrCP regulates the activity of both signalling pathways by mediating the ubiquitination of β-catenin and of the inhibitor of NF-κB, IκB (Hart et al., 1999; Jiang and Struhl, 1998; Latres et al., 1999; Spencer et al., 1999; Suzuki et al., 1999; Tan et al., 1999; Winston et al., 1999; Yaron et al., 1998). The Wnt/β-catenin and the canonical NF-κB signalling pathway play crucial roles in many key cellular processes such as cell proliferation and differentiation, and are often constitutively activated in various types of human cancer (Ben-Neriah and Karin; Bhatia et al., 2002; Clevers, 2006; Clevers and Nusse, 2012; DiDonato et al., 2012; Giles et al., 2003; Hayden and Ghosh, 2012; Inestrosa and Arenas, 2010; Klaus and Birchmeier, 2008; MacDonald et al., 2009; Merrill, 2012; Willert and Nusse, 2012). Thus, a comprehensive understanding of the regulation of Wnt/β-catenin and canonical NF-κB signalling is of academic as well as pharmacologic interest. In this thesis, the current biological knowledge regarding β-TrCP was collected and integrated into a mathematical framework of canonical NF-κB and Wnt/β-catenin signalling.

The study was based on two established detailed ODE models, one describing Wnt/β-catenin signalling (Lee et al., 2003) and one describing canonical NF-κB signalling (Lipniacki et al., 2004). The two detailed models were chosen as starting points because they have been well validated by experiments. However, the models had to be modified since they originally did not consider β-TrCP. To account for β-TrCP in the detailed model of canonical NF-κB signalling, two parameters were modified (Section 3.1). In the case of the detailed model of Wnt/β-catenin signalling, the model was extended by two transcriptional feedback mechanisms acting via the β-TrCP paralogues HOS and FWD1 (Section 4.1). The detailed models were used to explore the potential impact of transcriptional regulation of β-TrCP abundance on the dynamics of canonical NF-κB signalling and of Wnt/β-catenin signalling (Chapter 3 and Chapter 4, respectively).

Beside the effects of transcriptional variation of β-TrCP abundance, also the impact of competitive sequestration of β-TrCP by the canonical NF-κB and Wnt/β-catenin signalling pathway was investigated (Chapter 5). The analyses focussed in particular on the conditions that would enable or prevent mutual interaction (i.e. crosstalk) between both pathways via competitive β-TrCP sequestration. Instead of linking the detailed models of Wnt/β-catenin and canonical NF-κB signalling via β-TrCP in a large overall model, a minimal model of competitive β-TrCP sequestration was developed (Section 5.1). This was done for two reasons. First, there are no comprehensive experimental time series data sets available concerning Wnt/β-catenin pathway components as well as canonical NF-κB pathway components and β-TrCP in one cell type to estimate all kinetic parameters of a large overall model. Second, a large overall model would be difficult to analyse due to its much larger parameter space compared to a minimal model of competitive β-TrCP sequestration. The minimal model of competitive β-TrCP sequestration included two reduced and simplified modules of Wnt/β-catenin and canonical NF-κB signalling that were linked via the dynamics of β-TrCP (Figure 5.1). The minimal model was parameterised such that these two reduced pathway modules reproduced the simulations of the detailed models of canonical NF-κB and Wnt/β-catenin signalling (Figure 5.3).

Although the two detailed models, which were the starting points for the investigation, were originally developed to quantitatively reproduce experimental data, the results of this thesis must not be considered to be quantitative. The currently available information on β-TrCP considered in this thesis is of qualitative nature and in part even presumptive. For instance, β-TrCP is considered to be expressed at low concentrations in cells and it is thought to be unstable (Fuchs et al., 1999; Fuchs et al., 2004). Despite these uncertainties, general insights in the regulation of Wnt/β-catenin and canonical NF-κB signalling by β-TrCP could still be gained by the modelling approaches. These were in particular:

i.   The opposite regulation of FWD1 and HOS expression by Wnt/β-catenin signalling resulted in distinct regulatory effects of the β-TrCP feedbacks on the dynamics of the β-catenin/TCF complex (Chapter 4). Their actual impact on Wnt/β-catenin signalling depended on the effective expression levels of each β-TrCP paralogue. Simulations demonstrated that Wnt/β-catenin signalling dynamics were primarily affected by the HOS feedback under wild type conditions (Figure 4.5 and Figure 4.6), while the FWD1 feedback established a protection mechanism against loss of HOS expression (Figure 4.8).

ii. Decreasing β-TrCP abundance decreased the stimulated steady state concentration of nuclear NF-κB (Figure 3.3A) as well as reduced the amplitude of the transient dynamics of nuclear NF-κB upon TNF stimulation (Figure 3.4A). In contrast, decreasing β-TrCP abundance increased the stimulated steady state concentration of β-catenin (Figure 5.5B).

iii. Low β-TrCP abundance was not sufficient to enable crosstalk via competitive β-TrCP sequestration (Figure 5.9B). The impact of additional TNF stimulation onto the response dynamics of β-catenin upon Wnt stimulation was rather regulated by the parameters associated with β-TrCP-mediated IκB degradation in combination with those of β-TrCP production and degradation (Figure 5.10A).

These main insights (i. – iii.) were discussed in Section 3.3, Section 4.4, and Section 5.4 in more detail. The following sections discuss how these theoretical insights may offer alternative explanations for several seemingly contradicting experimental observations.

6.1.1 Abundance of HOS and FWD1 influence the impact of regulatory feedbacks in Wnt/β-catenin signalling

Transcriptional regulation of the expression of the β-TrCP paralogues HOS and FWD1 by Wnt/β-catenin signalling realises two transcriptional feedback mechanisms in the Wnt/β-catenin pathway (Figure 4.1). The influence of each individual feedback on pathway dynamics could be easily distinguished in the modelling approach by modulating their respective feedback strength. The analysis demonstrated that the opposite regulation of FWD1 and HOS expression levels by Wnt/β-catenin signalling results in distinct regulatory effects of the two feedbacks on the dynamics of the β-catenin/TCF complex. The HOS feedback had a strong impact on Wnt/β-catenin signalling dynamics while the FWD1 feedback hardly influenced the β-catenin/TCF dynamics in wild type conditions (Sections 4.3.1 - 4.3.3). Simulations showed that the FWD1 feedback rather established a protection mechanism in cells that lost HOS expression (Section 4.3.4). The model analysis thus showed that the actual impact of each feedback strongly depends on the absolute expression levels of FWD1 and HOS.

The theoretical investigation offers alternative explanations for certain experimental observations and their interpretations. In particular, it has been observed that knock-down and knock-out experiments of HOS or FWD1 have hardly a detectable impact on murine phenotype (Guardavaccaro et al., 2003). It has been concluded that HOS and FWD1 are functional redundant in these mice. The model simulations however suggest alternative explanations for the general absence of abnormalities. A siRNA-mediated knock-down of HOS expression levels may be compensated by an up-regulation of FWD1 according to the proposed protection mechanism (Figure 4.8). The FWD1 knock-out effects, on the other hand, may be masked by generally higher expression levels of HOS as was shown in Figure 4.7 and proposed by (Guardavaccaro et al., 2003). Furthermore, the model analyses suggest that, the observed testicular phenotype in FWD1 knock-out mice (Guardavaccaro et al., 2003) may indicate low concentrations of HOS in this particular tissue. This prediction could be checked in experiments.

6.1.2 Abundance of HOS and FWD1 influence the impact of transcriptional crosstalk from Wnt/β-catenin to canonical NF-κB signalling

Simulations of the detailed kinetic model of canonical NF-κB signalling predicted that the reduction of β-TrCP abundance decreases the amplitude and steady state concentration of nuclear NF-κB upon TNF stimulation (Figure 3.4B and Figure 3.3A, respectively). As Wnt/β-catenin signalling reduces the expression of the β-TrCP paralogue HOS, it was speculated that Wnt/β-catenin signalling can perhaps prevent canonical NF-κB signalling through this transcriptional crosstalk (Section 3.3). This speculative interpretation of the model simulations of Figure 3.4B is substantiated by experimental observations that Wnt/β-catenin signalling reduced canonical NF-κB activation through the regulation of HOS abundance in mouse embryonic fibroblast NIH3T3 cells (Spiegelman et al., 2000; Spiegelman et al., 2002b). In general, the impact of Wnt/β-catenin signalling on canonical NF-κB signalling is still controversially discussed. On the one hand, Wnt/β-catenin signalling is reported to inhibit NF-κB activity in colon and breast cancer cells (Deng et al., 2002; Deng et al., 2004), on the other hand it is reported to induced NF-κB activity in PC12 rat-derived neural progenitor cells and human embryonic kidney cells (Bournat et al., 2000; Spiegelman et al., 2000). The mechanistic insights gained in the presented theoretical study suggest that these seemingly contradicting experimental observations may derive from different expression levels of the β-TrCP paralogues and their opposite regulation by the Wnt/β-catenin pathway. If, for instance, HOS levels dominate over FWD1 levels in a particular cell, Wnt/β-catenin signalling may overall decrease total β-TrCP expression levels, which potentially inhibits canonical NF-κB signalling (Figure 5.5B). In contrast, if FWD1 levels dominate over HOS levels in the total cellular β-TrCP pool, Wnt/β-catenin signalling may overall increase total β-TrCP abundance, resulting in an activation of canonical NF-κB signalling (Figure 5.5B).

6.1.3 Regulation of protein abundance in the canonical NF-κB and Wnt/β-catenin signalling pathway may influence their crosstalk

β-TrCP is shared between the canonical NF-κB and Wnt/β-catenin signalling pathway. Its availability to one signalling pathway may thus be reduced by its sequestration into the other. Consequently, the canonical NF-κB and Wnt/β-catenin signalling pathways may influence each other's dynamics, i.e. the pathways may crosstalk with each other. The analysis of the minimal model of competitive β-TrCP sequestration revealed conditions that may support or prevent such potential crosstalk. Simulations demonstrated that the parameters associated with β-TrCP-mediated IκB degradation in combination with those of β-TrCP production and degradation predominantly influence the crosstalk impact of additional TNF stimulation on Wnt-stimulus-induced β-catenin dynamics (Section 5.3.5 and Section 5.3.6). This implies that β-TrCP has to remain bound to NF-κB/IκB in the β-TrCP/NF-κB/IκB complex and that unbound β-TrCP must be slowly reproduced in order to observe crosstalk. Interestingly, low β-TrCP abundance was shown to be no sufficient condition to enable crosstalk via competitive β-TrCP sequestration (Figure 5.9B). This theoretical finding contradicts the current view that a limited concentration of β-TrCP due to low β-TrCP abundance is the mechanistic reason for crosstalk between the canonical NF-κB and Wnt/β-catenin signalling pathway (Fuchs et al., 2004).

Preliminary results of the analysis of the minimal model indicated an influence of the relative concentrations of the species in the Wnt/β-catenin signalling pathway compared to the species of the canonical NF-κB signalling pathway on the direction of crosstalk between both pathways (Section 5.5.3). That is, if the concentration of the β-TrCP/NF-κB/IκB complex was much larger than that of the DC/β-catenin/β-TrCP complex (Figure 5.12A and C), simultaneous Wnt and TNF stimulation could alter the dynamics of β-catenin in comparison to single Wnt stimulation (Figure 5.7A). In contrast, the dynamics of NF-κB showed no observable difference between single TNF and simultaneous TNF and Wnt stimulation (Figure 5.7B). On the other hand, if the concentrations of the β-TrCP/NF-κB/IκB complex and the DC/β-catenin/β-TrCP complex were comparable, simultaneous Wnt and TNF stimulation changed the dynamics of β-catenin as well as NF-κB in comparison to single stimulations (Figure 5.13). It was thus speculated in Section 5.5 that regulation of protein abundance could be a possible mechanism to prevent crosstalk between Wnt/β-catenin and canonical NF-κB signalling. Recently, experiments on the quantification of pathway components of the Wnt/β-catenin pathway in resting mammalian cells have shown that central components such as Axin, APC, and β-catenin are expressed in a highly cell-type-specific manner (Chen et al., 2010; Schwanhausser et al., 2011; Tan et al., 2012). These experimental observations in combination with the modelling results propose a cell-type-dependent realisation of crosstalk (i.e. its detectability and direction) between the canonical NF-κB and the Wnt/β-catenin signalling pathway. The theoretical analysis furthermore underlined the necessity of comprehensive time series data sets of one particular cell type, to specify the modelling results to experimentally testable predictions. Once cell-type-specific experimental data sets become available, the minimal model can be easily rescaled to more appropriate values and is thus a versatile tool in future investigations.

6.1.4 Modulation of β-TrCP abundance to regulate SCF$^{β-TrCP}$ complex activity

In a recent review, the significance of modulating β-TrCP abundance as a major mode of regulation of SCF$^{β-TrCP}$ activity in the degradation of IκB and β-catenin has been questioned. It was argued that the entire SCF$^{β-TrCP}$ complex acts as a single functional E3 ubiquitin ligase unit and therefore would need a balanced stoichiometry of its subunits to function properly (Kanarek and Ben-Neriah, 2012). As supportive example, it was quoted that over-expression of wild type FWD1 in mice had been reported to disrupt proper SCF formation. The resulting aberrant NF-κB and Wnt/β-catenin signalling was thought to cause the tumourigenic phenotype of these mice (Belaidouni et al., 2005; Kanarek and Ben-Neriah, 2012). Thus, it was concluded in the review that variations of β-TrCP concentrations would probably be no means to regulate NF-κB and Wnt/β-catenin signalling (Kanarek and Ben-Neriah, 2012). This reasoning is challenged by the results of Section 4.3.5. There, simulations demonstrated that artificial overexpression of β-TrCP compared to transcriptional regulation of β-TrCP expression had very different impact on the dynamics of Wnt/β-catenin signal transduction. This finding indicates that β-TrCP overexpression experiments may hardly be instructive regarding the potential significance of modulating β-TrCP abundance as a major mode of regulating SCF$^{β-TrCP}$ activity in the canonical NF-κB and Wnt/β-catenin signalling pathway. The mathematical models used in this thesis have not considered stoichiometric effects of E3 complex formation but rather assumed that the enzymatic activity of the SCF$^{β-TrCP}$ complex scales linearly with β-TrCP concentration. In a first approximation, this assumption may adequately reflect the situation in a cell. The derived models proved to be useful to explore the consequences of changing β-TrCP availability on canonical NF-κB and Wnt/β-catenin signalling. In future studies, these models can easily be extended to account for stoichiometric changes in the subunits of the SCF$^{β-TrCP}$ complex, provided that experimental data will be available.

In summary, the analyses in this thesis showed (i) that FWD1 and HOS are functionally not redundant in regulating the Wnt/β-catenin signalling pathway, (ii) that the reduction of β-TrCP abundance, potentially due to transcriptional regulation by Wnt/β-catenin signalling, diminishes the NF-κB response to TNF stimulation, and (iii) that crosstalk of canonical NF-κB signalling to Wnt/β-catenin signalling via competitive β-TrCP sequestration is possibly under the condition of slow β-TrCP production and degradation in combination with enhanced retention of β-TrCP in the canonical NF-κB signalling pathway.

6.2 β-TrCP – a promising pharmaceutical target?

Aberrant activation of the Wnt/β-catenin and the NF-κB signalling pathways is detected in several severe human diseases, for instance autoimmunity, neurodegenerative diseases, and cancer (Ben-Neriah and Karin; Hayden and Ghosh, 2012; MacDonald et al., 2009). Despite extensive research in the field of Wnt/β-catenin signalling, drugs that inhibit aberrant activation of Wnt/β-catenin signalling are not yet available for clinical use (Barker and Clevers, 2006). This is partially due to the lack of accessible enzyme targets in the pathway and the complexity of pathway regulation (Zimmerman et al., 2012). Current therapeutic interventions to inhibit NF-κB activity focus on the interference with IκB degradation using proteasome inhibitors, such as Bortezomib (Frankland-Searby and Bhaumik, 2012). However, as the proteasome is directly or indirectly involved in almost all cellular processes, it is evident that continuous inhibition of the proteasome is toxic. An alternative approach to inhibit NF-κB activity would be to directly inhibit the ubiquitination of IκB by β-TrCP.

β-TrCP has been proposed as promising pharmacological target to reduce aberrant activation of Wnt/β-catenin and NF-κB signalling due to its intrinsic specificity for β-catenin and IκB (DiDonato et al., 2012; Fuchs et al., 2004; Kanarek and Ben-Neriah, 2012). The model analyses in this thesis demonstrated that modulation of β-TrCP levels in the pathways has strong regulatory impact on the response dynamics of canonical NF-κB and Wnt/β-catenin signalling (see Chapter 3 and Chapter 4, respectively). In order to modulate the availability of β-TrCP for Wnt/β-catenin and NF-κB signalling, research has focussed on identifying molecules that affect the association of β-TrCP with its targets. A mutant form of IκB called "IκB super-repressor" was reported to be very effective in experiments (Wang et al., 1996). Considering that β-TrCP recognise the identical phosphorylated amino acid sequences of IκB and β-catenin (Frescas and Pagano, 2008; Fuchs et al., 2004; Yaron et al., 1997), drugs that target β-TrCP influence Wnt/β-catenin and NF-κB signalling simultaneously. Targeting β-TrCP in constitutive activated NF-κB signalling by inhibiting the association of β-TrCP with IκB also interferes with β-TrCP binding to β-catenin resulting in an increased activity of β-catenin. Vice versa, interference with β-TrCP to counteract aberrant activation of Wnt/β-catenin signalling leads to an activation of NF-κB signalling. Simulations in this thesis also demonstrated that the impact of β-TrCP modulation is not pathway-specific (Figure 5.5B).

To achieve pathway specificity of pharmaceutical targeting of β-TrCP, the undesirable side effects on the other pathway need to be omitted at the same time. One possibility is to target IκB instead of β-TrCP in the β-TrCP-IκB interaction. Using that strategy, the interaction of β-catenin and β-TrCP is not affected. An NF-κB pathway-specific small molecule inhibitor of IκBα ubiquitination (GS143) has already been found that apparently does not simultaneously promote the accumulation of β-catenin (Nakajima et al., 2008). Clinically applicable small molecule inhibitors that are specific for Wnt/β-catenin signalling have yet to be identified (Barker and Clevers, 2006).

The Wnt/β-catenin and the NF-κB signalling pathway are often constitutively activated in human cancer. Both pathways induce the expression of the proto-oncogenic transcription factor c-myc and the cell-cycle regulator cyclin D1 resulting in enhanced cell proliferation, a hallmark of cancer (Albanese et al., 2003; Archbold et al., 2012; Hanahan and Weinberg, 2000; Hanahan and Weinberg, 2011; Joyce et al., 2001). The molecular causes leading to constitutive NF-κB activation are in most cases not well understood since components of the signalling pathway are rarely mutated (Staudt, 2010). Considering the Wnt/β-catenin pathway, constitutive activation is in most cases caused by mutations in β-catenin itself or components of the destruction complex such as APC or Axin/Axin2 (Giles et al., 2003). These mutations have in common to interfere with the phosphorylation of β-catenin in the destruction complex leading to nuclear accumulation of β-catenin. A particular example is human colon cancer. In about 80% of the human colon cancer samples APC mutations are detected (Giles et al., 2003). The APC mutations interfere with the phosphorylation of β-catenin in the destruction complex such that β-TrCP cannot recognise β-catenin to mediate its ubiquitination and degradation. Simulations of APC mutations (see Section 2.3) using the detailed HOS and FWD1 feedback model of Wnt/β-catenin signalling showed that the mutation-induced increase of the steady state concentration of β-catenin is paralleled by a strong up-regulation of the expression level of FWD1 (Figure 4.10). This theoretical finding agrees well with experimental observations in these cancer types (Frescas and Pagano, 2008; Fuchs et al., 2004; Saitoh and Katoh, 2001). The simulations demonstrated that even strongly up-regulated β-TrCP concentrations cannot decrease the elevated concentrations of β-catenin. Thus, it is reasonable to assume that pharmaceutical targeting of β-TrCP is futile to control the aberrant activation of the Wnt/β-catenin pathway in colon cancer. The up-regulation of β-TrCP can also increase NF-κB signalling, which was shown in simulations (Figure 5.5B) and observed in experiments (Kroll et al., 1999; Muerkoster et al., 2005; Spiegelman et al., 2000; Spiegelman et al., 2001). Up-regulated NF-κB activity often promotes cell resistance to anti-cancer therapy (Bhatia et al., 2002; DiDonato et al., 2012) and even induces the secretion of the tumour-promoting cytokines interleukin-6 and TNFα (Karin, 2009). In this regard, pharmacological targeting of β-TrCP could be useful to inhibit NF-κB activity despite its ineffectiveness in the regulation of the aberrant activation of Wnt/β-catenin signalling.

Taken together, direct pharmacological targeting of β-TrCP is problematic due to its involvement in different signalling pathways. Pathway-specific interference however may be a promising approach. This can either be achieved by pathway-specific drugs, such as the small molecule inhibitor of IκBα ubiquitination (GS143) (Nakajima et al., 2008), or by targeting β-TrCP in particular cellular conditions, such as in colon cancer cells.

6.3   Outlook on future investigations

A few issues remain unresolved in the thesis. For instance, possible mechanisms of pathway insulation in the minimal model of competitive β-TrCP sequestration were merely sketched in Section 5.5. Even in the absence of experimental data, it would be very interesting to continue this investigation in a mathematical modelling approach. The minimal model also misses the transcriptional regulation of β-TrCP by Wnt/β-catenin signalling. A minimal model extended by the HOS and FWD1 feedback mechanisms may be useful in a more thorough exploration of bifurcations than performed for the two-feedback model (Appendix 7.4.5) due to the great reduction of parameter space compared to the detailed model. Possibly, this minimal model approach may also guide towards identifying mechanistic reasons for the very long period of the limit cycles observed in the two-feedback model (Figure 8.2). Furthermore, an inclusion of the HOS and FWD1 feedback mechanisms into the minimal model establishes dependence between β-catenin and NF-κB at stimulated steady state condition that may create additional crosstalk between both pathways in the model.

6.3.1 Additional aspects connected to β-TrCP in Wnt/β-catenin and NF-κB signalling

This thesis has been focusing on only two β-TrCP targets, i.e. β-catenin and IκB (more precisely, the IκBα isoform). However, β-TrCP was also shown to control the degradation of IκBβ, IκBε, and p105 as well as to mediate the processing of p105 to p50 (Heissmeyer et al., 2001; Lang et al., 2003; Orian et al., 2000; Shirane et al., 1999; Wu and Ghosh, 1999). Competition of these pathway components for β-TrCP binding may play a role in the coordination of these different IκB isoforms and their feedbacks. In experiments, distinct degradation kinetics have been observed for the different IκBs (Hinz et al., 2012). The distinct kinetics may create a specific temporal demand for β-TrCP availability by each IκB isoform. The IκB isoforms could thus become kinetically insulated (see Section 5.5) with respect to β-TrCP despite their common control by β-TrCP. To explore potential coordination between the IκB isoforms in future studies, sensitivity analysis and bifurcation analysis (as done in Chapter 3) of further detailed mathematical models of NF-κB signalling (Hoffmann et al., 2002; Shih et al., 2009; Werner et al., 2005) may be a promising approach.

β-TrCP was also reported to mediate the processing of p100 to p52 in the non-canonical NF-κB signalling pathway (Fong and Sun, 2002). Since non-canonical NF-κB signalling is responding to stimulation in the time frame of hours (Scheidereit, 2006), quite similar to the Wnt/β-catenin signalling pathway (Luckert et al., 2012), it is tempting to speculate on crosstalk between the non-canonical NF-κB and the Wnt/β-catenin pathway via β-TrCP. As both signalling pathways are involved in developmental processes (Dominguez et al., 2009; Huntzicker and Oro, 2008), crosstalk may possibly be of importance in the coordination of these processes. The minimal modelling approach of competitive β-TrCP sequestration used in this thesis may prove useful to gain fundamental insights into the potential molecular mechanisms of coordination between the non-canonical NF-κB and the Wnt/β-catenin pathway.

6.3.2 Additional potential mechanisms of crosstalk between Wnt/β-catenin and NF-κB signalling

The impact that Wnt/β-catenin and NF-κB signalling have on each other has been object of experimental research for years and is still controversially discussed. For instance, β-catenin was reported to induce NF-κB activity in human embryonic kidney cells and PC12 rat-derived neural progenitor cells (Bournat et al., 2000; Spiegelman et al., 2000). In contrast, β-catenin was shown to inhibit NF-κB activity in human colon cancer as well as breast cancer cell lines (Deng et al., 2002; Deng et al., 2004). The present thesis focused on crosstalk mechanisms between the Wnt/β-catenin and canonical NF-κB signalling pathway via β-TrCP since β-TrCP represents a direct molecular link between β-catenin and NF-κB activation. Further possible crosstalk mechanisms between the Wnt/β-catenin and the NF-κB signalling pathway have been proposed in the literature. For instance, the leucine zipper putative tumour suppressor 2 (LZTS2) regulates nuclear export of β-catenin and was shown to repress the transcriptional activity of β-catenin if overexpressed (Thyssen et al., 2006). LZTS2 is transcriptionally regulated by NF-κB in a cell-type-specific manner. NF-κB is reported to inhibit the transcription of LZTS2 in glioma cancer cells (Cho et al., 2008) and human adipose tissue-derived mesenchymal stem cells (Hyun Hwa et al., 2008), but to increase the expression of LZTS2 in cell lines of colon, breast, and hepatic cancer (Cho et al., 2008). Another crosstalk mechanism involves phosphorylation of β-catenin by the kinases IKKα and IKKβ. It was reported that IKKα-dependent phosphorylation increases whereas IKKβ-dependent phosphorylation of β-catenin decreases the transcriptional activity of β-catenin (Lamberti et al., 2001). In addition, IKKα was shown to stabilise β-catenin and stimulate β-catenin/TCF complex-dependent transcription through an NF-κB-independent mechanism (Carayol and Wang, 2006). A third potential crosstalk mechanism focuses on GSK3, which is part of the β-catenin destruction complex. GSK3 was shown to modulate IKK complex activity as well as to change the protein-protein-interaction network of the NF-κB subunit p65 and consequently the transcriptional activity of NF-κB (reviewed in (Billadeau, 2007)). Yet, the particular impact of GSK3 on the transcriptional activity of NF-κB remains controversial, with reports suggesting NF-κB activation (Hoeflich et al., 2000; Schwabe and Brenner, 2002) and others reporting NF-κB inhibition by GSK3 activity (Bournat et al., 2000; Gotschel et al., 2008; Sanchez et al., 2003). Importantly, the experiments that interfered with GSK activity were performed using inhibitors, siRNAs or overexpression of GSK3, instead of applying

Wnt ligands (Bournat et al., 2000; Gotschel et al., 2008; Hoeflich et al., 2000; Sanchez et al., 2003; Schwabe and Brenner, 2002). Consequently, the direct proof that Wnt/β-catenin signalling influences NF-κB signalling via GSK3 has yet to be presented. In addition, it was shown that only the fraction of GSK3 molecules that is bound by the destruction complex is relevant in Wnt/β-catenin signalling (McNeill and Woodgett, 2010). The application of inhibitors, siRNAs or overexpression of GSK3 however affects total GSK3 in the cell. Consequently, the contradicting impact on NF-κB activity may as well be due to other mechanisms than GSK3-dependent crosstalk between Wnt/β-catenin and NF-κB signalling.

Interactions between the Wnt/β-catenin and NF-κB signalling pathways have furthermore been reported on DNA level. β-Catenin can directly bind to the p65 as well as the p50 subunit of NF-κB resulting in reduced NF-κB-DNA-binding. β-Catenin thus interferes with NF-κB target gene expression in several cell types, such as human colon cancer cells, mouse embryonic fibroblasts, and hepatocytes (Deng et al., 2002; Duan et al., 2007; Nejak-Bowen et al., 2012; Sun et al., 2005). Wnt/β-catenin and NF-κB signalling may also synergistically co-regulate target genes such as Lef1 and cyclin D1 expression (Albanese et al., 2003; Yun et al., 2007).

These examples of various crosstalk mechanisms on different levels of signal transduction indicate the potentially complex interdependence of the Wnt/β-catenin and NF-κB signalling pathway. This complexity of the interaction network renders it very difficult to understand how Wnt and TNF signals are transduced through the network to regulate the expression of target genes. Mathematical modelling approaches constitute a sophisticated method to purposefully address this complexity. Especially, the models described in this thesis may provide an advantageous starting point for future investigations into the Wnt/β-catenin and NF-κB signalling pathway network.

Appendix

7. Mathematical models

7.1 The detailed kinetic model of canonical NF-κB signalling

In Section 2.1, the schematic representation of the detailed kinetic model of canonical NF-κB signalling is shown (Figure 2.1). In the scheme, the numbers next to arrow denote the numbers of the particular reactions. The rate equations that describe the reactions are provided in Section 7.1.3. In the rate equations, components in a complex are separated by a slash. The model is given by the Equations 7.1 - 7.44. The parameters are listed in Table 7.1.

7.1.1 Differential equations

$$\frac{d(IKKn)}{dt} = v_1 - v_2 - v_3 \tag{7.1}$$

$$\frac{d(IKKi)}{dt} = v_4 - v_5 + v_{26} \tag{7.2}$$

$$\frac{d(IKKa)}{dt} = v_3 - v_4 - v_6 - v_7 + v_8 - v_9 + v_{10} - v_{26} \tag{7.3}$$

$$\frac{d(IKKa/I\kappa B)}{dt} = v_7 - v_8 \tag{7.4}$$

$$\frac{d(IKKa/I\kappa B/NF\text{-}\kappa B)}{dt} = v_9 - v_{10} \tag{7.5}$$

$$\frac{d(I\kappa B)}{dt} = -v_7 - v_{14} + v_{16} - v_{17} + v_{20} - v_{21} \tag{7.6}$$

$$\frac{d(NF\text{-}\kappa B)}{dt} = v_{10} - v_{11} - v_{14} + v_{15} \tag{7.7}$$

$$\frac{d(I\kappa Bnuc)}{dt} = -v_{12} - k_v \cdot v_{16} + k_v \cdot v_{17} \tag{7.8}$$

$$\frac{d(NF\text{-}\kappa Bnuc)}{dt} = k_v \cdot v_{11} - v_{12} \qquad\qquad 7.9$$

$$\frac{d(I\kappa Bnuc/NF\text{-}\kappa Bnuc)}{dt} = v_{12} - k_v \cdot v_{13} \qquad\qquad 7.10$$

$$\frac{d(I\kappa B\text{-}mRNA)}{dt} = v_{18} - v_{19} \qquad\qquad 7.11$$

$$\frac{d(A20\text{-}mRNA)}{dt} = v_{22} - v_{23} \qquad\qquad 7.12$$

$$\frac{d(A20)}{dt} = v_{24} - v_{25} \qquad\qquad 7.13$$

$$\frac{d(cgen\text{-}mRNA)}{dt} = v_{27} - v_{28} \qquad\qquad 7.14$$

7.1.2 Conservation relation

$$(NF\text{-}\kappa B)_{total} = (IKKa/I\kappa B/NF\text{-}\kappa B) + (NF\text{-}\kappa B) + (I\kappa B/NF\text{-}\kappa B) + \frac{1}{k_v} \cdot (NF\text{-}\kappa Bnuc)$$

$$+ \frac{1}{k_v} \cdot (I\kappa Bnuc/NF\text{-}\kappa Bnuc) \qquad\qquad 7.15$$

7.1.3 Rate equations

$$v_1 = constant \qquad\qquad 7.16$$

$$v_2 = k_2 \cdot (IKKn) \qquad\qquad 7.17$$

$$v_3 = k_3 \cdot TNF \cdot (IKKn) \qquad\qquad 7.18$$

$$v_4 = k_4 \cdot (IKKa) \qquad\qquad 7.19$$

$$v_5 = k_5 \cdot (IKKi) \qquad\qquad 7.20$$

$$v_6 = k_6 \cdot (IKKa) \qquad\qquad 7.21$$

$$v_7 = k_7 \cdot (IKKa) \cdot (I\kappa B) \qquad\qquad 7.22$$

$$v_8 = k_8 \cdot (IKKa/I\kappa B) \qquad\qquad 7.23$$

$$v_9 = k_9 \cdot (IKKa) \cdot (I\kappa B/NF\text{-}\kappa B) \qquad\qquad 7.24$$

$$v_{10} = k_{10} \cdot (IKKa/I\kappa B/NF\text{-}\kappa B) \qquad\qquad 7.25$$

$$v_{11} = k_{11} \cdot (NF\text{-}\kappa B) \qquad\qquad 7.26$$

$$v_{12} = k_{12} \cdot (I\kappa Bnuc) \cdot (NF\text{-}\kappa Bnuc) \qquad\qquad 7.27$$

$$v_{13} = k_{13} \cdot (I\kappa Bnuc/NF\text{-}\kappa Bnuc) \qquad\qquad 7.28$$

$$v_{14} = k_{14} \cdot (I\kappa B) \cdot (NF\text{-}\kappa B) \qquad\qquad 7.29$$

$$v_{15} = k_{15} \cdot (I\kappa B/NF\text{-}\kappa B) \qquad\qquad 7.30$$

$$v_{16} = k_{16} \cdot (I\kappa Bnuc) \qquad\qquad 7.31$$

$$v_{17} = k_{17} \cdot (I\kappa B) \qquad\qquad 7.32$$

$$v_{18} = k_{18} \cdot (NF\text{-}\kappa Bnuc) \qquad\qquad 7.33$$

$$v_{19} = k_{19} \cdot (I\kappa B\text{-}mRNA) \qquad\qquad 7.34$$

$$v_{20} = k_{20} \cdot (I\kappa B\text{-}mRNA) \qquad\qquad 7.35$$

$$v_{21} = k_{21} \cdot (I\kappa B) \qquad\qquad 7.36$$

$$v_{22} = k_{22} \cdot (NF\text{-}\kappa Bnuc) \qquad\qquad 7.37$$

$$v_{23} = k_{23} \cdot (A20\text{-}mRNA) \qquad\qquad 7.38$$

$$v_{24} = k_{24} \cdot (A20\text{-}mRNA) \qquad\qquad 7.39$$

$$v_{25} = k_{25} \cdot (A20) \qquad\qquad 7.40$$

$$v_{26} = k_{26} \cdot TNF \cdot (IKKa) \cdot (A20) \qquad\qquad 7.41$$

$$v_{27} = k_{27} \cdot (NF\text{-}\kappa Bnuc) \qquad\qquad 7.42$$

$$v_{28} = k_{28} \cdot (cgen\text{-}mRNA) \qquad\qquad 7.43$$

7.1.4   Stimulation by TNF

The TNF stimulus is simulated by incrementing TNF in Equation 7.18 and Equation 7.41 from 0 to 1 at the time point of stimulus application.

$$TNF_{t=0} = 0 \text{ and } TNF_{t>0} = 1 \qquad\qquad 7.44$$

7.1.5 Model parameters

Table 7.1: Parameters of the detailed kinetic model of canonical NF-κB signalling.
The parameter values are adopted form (Lipniacki et al., 2004).

Parameter	Value	
v_1	1.5	$nM \cdot min^{-1}$
k_2	$7.5 \cdot 10^{-3}$	min^{-1}
k_3	0.15	min^{-1}
k_4	$9 \cdot 10^{-2}$	min^{-1}
k_5	$7.5 \cdot 10^{-3}$	min^{-1}
k_6	$7.5 \cdot 10^{-3}$	min^{-1}
k_7	$1.2 \cdot 10^{-2}$	$nM^{-1} \cdot min^{-1}$
k_8	6	min^{-1}
k_9	$6 \cdot 10^{-2}$	$nM^{-1} \cdot min^{-1}$
k_{10}	6	min^{-1}
k_{11}	0.15	min^{-1}
k_{12}	$3 \cdot 10^{-2}$	$nM^{-1} \cdot min^{-1}$
k_{13}	0.6	min^{-1}
k_{14}	$3 \cdot 10^{-2}$	$nM^{-1} \cdot min^{-1}$
k_{15}	$1.2 \cdot 10^{-3}$	min^{-1}
k_{16}	$3 \cdot 10^{-2}$	min^{-1}
k_{17}	$6 \cdot 10^{-2}$	min^{-1}
k_{18}	$3 \cdot 10^{-5}$	min^{-1}
k_{19}	$2.4 \cdot 10^{-2}$	min^{-1}

Table 7.1 continues at the next page.

Table 7.1: Parameters of the detailed kinetic model of canonical NF-κB signalling. (continued)

The parameter values are adopted form (Lipniacki et al., 2004).

Parameter	Value	
k_{20}	30	min^{-1}
k_{21}	$6 \cdot 10^{-3}$	min^{-1}
k_{22}	$3 \cdot 10^{-5}$	min^{-1}
k_{23}	$2.4 \cdot 10^{-2}$	min^{-1}
k_{24}	30	min^{-1}
k_{25}	$1.8 \cdot 10^{-2}$	min^{-1}
k_{26}	$6 \cdot 10^{-3}$	$\text{nM}^{-1} \cdot \text{min}^{-1}$
k_{27}	$3 \cdot 10^{-5}$	min^{-1}
k_{28}	$2.4 \cdot 10^{-2}$	min^{-1}
$(\text{NF-κB})_{total}$	60	nM
k_v	5	
$k_{\beta\text{-TrCP}}$	6	min^{-1}

Note that the rate constants (k_8 and k_{10}) of the two phosphorylation-dependent IκB degradation reactions (Equation 7.23 and Equation 7.25, respectively) are substituted by the rate constant $k_{\beta\text{-TrCP}}$ in the simulations of Chapter 3.

7.2 The detailed model of Wnt/β-catenin signalling

In Section 2.2, the schematic representation of the detailed kinetic model of Wnt/β-catenin signalling is shown (Figure 2.2). In the scheme, the numbers next to arrow denote the numbers of the particular reactions. The rate equations that describe the reactions are provided in Section 7.2.4. In the model, components in a complex are separated by a slash. The asterisk symbolizes posttranslational modifications. The model is given by the Equations 7.45 - 7.83. The set of differential Equations 7.45 - 7.59 is simplified to eight algebraic and seven differential equations according to the supplementary information of reference (Lee et al., 2003), which considers in its approach the conservation relations and rapid equilibrium approximations of binding reactions (Section 7.2.2 and Section 7.2.3, respectively). The parameters are listed in Table 7.3, Table 7.4, and Table 7.5.

7.2.1 Differential equations

$$\frac{d(Dsh_a)}{dt} = v_1 - v_2 \tag{7.45}$$

$$\frac{d(Dsh_i)}{dt} = -v_1 + v_2 \tag{7.46}$$

$$\frac{d(GSK3)}{dt} = 0 \tag{7.47}$$

$$\frac{d(Axin)}{dt} = -v_7 + v_{14} - v_{15} \tag{7.48}$$

$$\frac{d(APC)}{dt} = -v_7 - v_{17} \tag{7.49}$$

$$\frac{d(APC/Axin)}{dt} = v_3 - v_6 + v_7 \tag{7.50}$$

$$\frac{d(APC/Axin/GSK3)}{dt} = -v_3 - v_4 + v_5 + v_6 \tag{7.51}$$

$$\frac{d(APC^*/Axin^*/GSK3)}{dt} = v_4 - v_5 - v_8 + v_{10} \tag{7.52}$$

$$\frac{d(APC^*/Axin^*/GSK3/\beta\text{-}catenin)}{dt} = v_8 - v_9 \tag{7.53}$$

$$\frac{d(APC^*/Axin^*/GSK3/\beta\text{-}catenin^*)}{dt} = v_9 - v_{10} \tag{7.54}$$

$$\frac{d(\beta\text{-}catenin)}{dt} = -v_8 + v_{12} - v_{13} - v_{16} - v_{17} \qquad\qquad 7.55$$

$$\frac{d(\beta\text{-}catenin^*)}{dt} = v_{10} - v_{11} \qquad\qquad 7.56$$

$$\frac{d(TCF)}{dt} = -v_{16} \qquad\qquad 7.57$$

$$\frac{d(\beta\text{-}catenin/TCF)}{dt} = v_{16} \qquad\qquad 7.58$$

$$\frac{d(APC/\beta\text{-}catenin)}{dt} = v_{17} \qquad\qquad 7.59$$

7.2.2 Conservation relations

$$Dsh_{total} = (Dsh_a) + (Dsh_i) \qquad\qquad 7.60$$

$$GSK3_{total} = (GSK3) \qquad\qquad 7.61$$

$$APC_{total} = (APC) + (APC/\beta\text{-}catenin) \qquad\qquad 7.62$$

$$TCF_{total} = (TCF) + (\beta\text{-}catenin/TCF) \qquad\qquad 7.63$$

7.2.3 Binding equilibria

$$K_7 = \frac{(APC) \cdot (Axin)}{(APC/Axin)} \qquad\qquad 7.64$$

$$K_8 = \frac{(APC^*/Axin^*/GSK3) \cdot (\beta\text{-}catenin)}{(APC^*/Axin^*/GSK3/\beta\text{-}catenin)} \qquad\qquad 7.65$$

$$K_{16} = \frac{(TCF) \cdot (\beta\text{-}catenin)}{(\beta\text{-}catenin/TCF)} \qquad\qquad 7.66$$

$$K_{17} = \frac{(APC) \cdot (\beta\text{-}catenin)}{(APC/\beta\text{-}catenin)} \qquad\qquad 7.67$$

7.2.4 Rate equations

$$v_1 = k_1 \cdot Wnt \cdot (Dsh_i) \qquad\qquad 7.68$$

$$v_2 = k_2 \cdot (Dsh_a) \qquad\qquad 7.69$$

$$v_3 = k_3 \cdot (Dsh_a) \cdot (APC/Axin/GSK3) \qquad\qquad 7.70$$

$$v_4 = k_4 \cdot (APC/Axin/GSK3) \tag{7.71}$$

$$v_5 = k_5 \cdot (APC^*/Axin^*/GSK3) \tag{7.72}$$

$$v_6 = k_6 \cdot (GSK3) \cdot (APC/Axin) - k_{6r} \cdot (APC/Axin/GSK3) \tag{7.73}$$

$$v_9 = k_9 \cdot (APC^*/Axin^*/GSK3/\beta\text{-}catenin) \tag{7.74}$$

$$v_{10} = k_{10} \cdot (APC^*/Axin^*/GSK3/\beta\text{-}catenin^*) \tag{7.75}$$

$$v_{11} = k_{11} \cdot (\beta\text{-}catenin^*) \tag{7.76}$$

$$v_{12} = constant \tag{7.77}$$

$$v_{13} = k_{13} \cdot (\beta\text{-}catenin) \tag{7.78}$$

$$v_{14} = constant \tag{7.79}$$

$$v_{15} = k_{15} \cdot (Axin) \tag{7.80}$$

7.2.5 Stimulation by Wnt

Absence of Wnt (unstimulated): $\quad Wnt = 0 \tag{7.81}$

Constant Wnt stimulation: $\quad Wnt_{t=0} = 0$ and $Wnt_{t>0} = 1 \tag{7.82}$

Transient Wnt stimulation: $\quad Wnt_{t=0} = 0$ and $Wnt_{t>0} = Exp[-\lambda \cdot t] \tag{7.83}$

Table 7.2: Parameter of the transient Wnt stimulation.

The parameter value is adopted form (Lee et al., 2003).

Parameter	Value	
λ	0.05	min^{-1}

7.2.6 Model parameters

Table 7.3: Parameters of the detailed Wnt/β-catenin model.

The parameter values are adopted form (Lee et al., 2003).

Parameter	Value	
k_1	0.182	min^{-1}
k_2	$1.82 \cdot 10^{-2}$	min^{-1}
k_3	$5 \cdot 10^{-2}$	$nM^{-1} \cdot min^{-1}$
k_4	0.267	min^{-1}
k_5	0.133	min^{-1}
k_6	$9.09 \cdot 10^{-2}$	$nM^{-1} \cdot min^{-1}$
k_{6r}	0.909	min^{-1}
k_9	$2.06 \cdot 10^{2}$	min^{-1}
k_{10}	$2.06 \cdot 10^{2}$	min^{-1}
k_{11}	0.417	min^{-1}
v_{12}	0.423	$nM \cdot min^{-1}$
k_{13}	$2.57 \cdot 10^{-4}$	min^{-1}
v_{14}	$8.22 \cdot 10^{-5}$	$nM \cdot min^{-1}$
k_{15}	0.167	min^{-1}

Table 7.4: Dissociation constants.

The parameter values are adopted form (Lee et al., 2003).

Parameter	Value	
K_7	50	nM
K_8	120	nM
K_{16}	30	nM
K_{17}	$1.2 \cdot 10^{-3}$	nM

Table 7.5: Total protein concentrations of proteins obeying a conservation relation.

The parameter values are adopted form (Lee et al., 2003).

Component	Value	
APC_{total}	100	nM
GSK_{total}	50	nM
TCF_{total}	15	nM
Dsh_{total}	100	nM

7.3 The two-feedback model of Wnt/β-catenin signalling including the FWD1 and HOS feedback

In Chapter 4, the schematic representation of the two-feedback model is shown (Figure 4.1). In the scheme, the reactions of the model marked in black are adapted from the detailed model of Wnt/β-catenin signalling (Lee et al., 2003) (see Section 2.2). These reactions are extended by the HOS and FWD1 feedback mechanisms (green and blue, respectively), the ubiquitination reactions (red), and the TCF turnover mechanism (grey). In the scheme, the number next to arrow denotes the number of the particular reaction. The rate equations that describe the reactions are provided in Section 7.3.3. In the rate equations, components in a complex are separated by a slash and the asterisk denotes phosphorylation. The two-feedback model is given by the Equations 7.84 - 7.143. The parameters of Table 7.7 and Table 7.8 define the wild type scenario. As in the case of wild type cells, Equations 7.84 - 7.143 are also used to model APC mutant cells. The parameters used to simulate the APC mutations are given in Table 7.7 and Table 7.8 with the exception of the parameters of Equation 7.113, Equation 7.114, and Equation 7.124. These parameters (Table 7.9) are adopted from (Cho et al., 2006) to account for the changes in binding efficiency of β-catenin and/or Axin to APC due to APC mutation (see Section 2.3).

7.3.1 Differential equations

$$\frac{d(Dsh_a)}{dt} = v_1 - v_2 \qquad\qquad 7.84$$

$$\frac{d(GSK3)}{dt} = v_3 - v_6 \qquad\qquad 7.85$$

$$\frac{d(Axin)}{dt} = -v_7 + v_{14} - v_{15} \qquad\qquad 7.86$$

$$\frac{d(APC)}{dt} = -v_7 - v_{17} \qquad\qquad 7.87$$

$$\frac{d(APC/Axin)}{dt} = v_3 - v_6 + v_7 \qquad\qquad 7.88$$

$$\frac{d(APC/Axin/GSK3)}{dt} = -v_3 - v_4 + v_5 + v_6 \qquad\qquad 7.89$$

$$\frac{d(APC^*/Axin^*/GSK3)}{dt} = v_4 - v_5 - v_8 + v_{10} + v_{10b} \qquad\qquad 7.90$$

$$\frac{d(APC^*/Axin^*/GSK3/\beta\text{-}catenin)}{dt} = v_8 - v_9 \qquad 7.91$$

$$\frac{d(APC^*/Axin^*/GSK3/\beta\text{-}catenin^*)}{dt} = v_9 - v_{22} - v_{22b} \qquad 7.92$$

$$\frac{d(\beta\text{-}catenin)}{dt} = -v_8 + v_{12} - v_{13} - v_{16} - v_{17} \qquad 7.93$$

$$\frac{d(\beta\text{-}catenin^{*/ub})}{dt} = v_{10} + v_{10b} - v_{11} \qquad 7.94$$

$$\frac{d(TCF)}{dt} = -v_{16} + v_{50} - v_{51} \qquad 7.95$$

$$\frac{d(\beta\text{-}catenin/TCF)}{dt} = v_{16} \qquad 7.96$$

$$\frac{d(APC/\beta\text{-}catenin)}{dt} = v_{17} \qquad 7.97$$

$$\frac{d(HOS\ mRNA)}{dt} = v_{18} - v_{19} \qquad 7.98$$

$$\frac{d(HOS)}{dt} = v_{10} + v_{20} - v_{21} - v_{22} \qquad 7.99$$

$$\frac{d(CRD\text{-}BP\ mRNA)}{dt} = v_{23} - v_{24} \qquad 7.100$$

$$\frac{d(CRD\text{-}BP)}{dt} = v_{25} - v_{26} \qquad 7.101$$

$$\frac{d(FWD1\ mRNA)}{dt} = -v_{27} + v_{28} \qquad 7.102$$

$$\frac{d(FWD1)}{dt} = v_{10b} - v_{22b} + v_{29} - v_{30} \qquad 7.103$$

7.3.2 Conservation relations

$$Dsh_{total} = (Dsh_a) + (Dsh_i) \qquad 7.104$$

$$GSK3_{total} = (GSK3) + (APC/Axin/GSK3) + (APC^*/Axin^*/GSK3)$$

$$+(APC^*/Axin^*/GSK3/\beta\text{-}catenin) + (APC^*/Axin^*/GSK3/\beta\text{-}catenin^*)$$

$$+(APC^*/Axin^*/GSK3/\beta\text{-}catenin^*/HOS)$$

$$+(APC^*/Axin^*/GSK3/\beta\text{-}catenin^*/FWD1) \qquad 7.105$$

$$APC_{total} = (APC) + (APC/Axin) + (APC/Axin/GSK3) + (APC^*/Axin^*/GSK3)$$

$$+(APC^*/Axin^*/GSK3/\beta\text{-}catenin) + (APC^*/Axin^*/GSK3/\beta\text{-}catenin^*)$$

$$+(APC^*/Axin^*/GSK3/\beta\text{-}catenin^*/HOS)$$

$$+(APC^*/Axin^*/GSK3/\beta\text{-}catenin^*/FWD1) + (APC/\beta\text{-}catenin) \qquad 7.106$$

7.3.3 Rate equations

$$v_1 = k_1 \cdot Wnt \cdot (Dsh_i) \qquad\qquad 7.107$$

$$v_2 = k_2 \cdot (Dsh_a) \qquad\qquad 7.108$$

$$v_3 = k_3 \cdot (Dsh_a) \cdot (APC/Axin/GSK3) \qquad\qquad 7.109$$

$$v_4 = k_4 \cdot (APC/Axin/GSK3) \qquad\qquad 7.110$$

$$v_5 = k_5 \cdot (APC^*/Axin^*/GSK3) \qquad\qquad 7.111$$

$$v_6 = k_6 \cdot (GSK3) \cdot (APC/Axin) - k_{6r} \cdot (APC/Axin/GSK3) \qquad\qquad 7.112$$

$$v_7 = k_7 \cdot (APC) \cdot (Axin) - k_{7r} \cdot (APC/Axin) \qquad\qquad 7.113$$

$$v_8 = k_8 \cdot (APC^*/Axin^*/GSK3) \cdot (\beta\text{-}catenin) - k_{8r} \cdot (APC^*/Axin^*/GSK3/\beta\text{-}catenin) \qquad 7.114$$

$$v_9 = k_9 \cdot (APC^*/Axin^*/GSK3/\beta\text{-}catenin) - k_{9r} \cdot (APC^*/Axin^*/GSK3/\beta\text{-}catenin^*) \qquad 7.115$$

$$v_{10} = k_{10} \cdot (APC^*/Axin^*/GSK3/\beta\text{-}catenin^*/HOS) \qquad\qquad 7.116$$

$$v_{10b} = k_{10b} \cdot (APC^*/Axin^*/GSK3/\beta\text{-}catenin^*/FWD1) \qquad\qquad 7.117$$

$$v_{11} = k_{11} \cdot (\beta\text{-}catenin^{*/ub}) \qquad\qquad 7.118$$

$$v_{12} = constant \qquad\qquad 7.119$$

$$v_{13} = k_{13} \cdot (\beta\text{-}catenin) \qquad\qquad 7.120$$

$$v_{14} = constant \qquad\qquad 7.121$$

$$v_{15} = k_{15} \cdot (Axin) \qquad\qquad 7.122$$

$$v_{16} = k_{16} \cdot (TCF) \cdot (\beta\text{-}catenin) - k_{16r} \cdot (\beta\text{-}catenin/TCF) \qquad\qquad 7.123$$

$$v_{17} = k_{17} \cdot (APC) \cdot (\beta\text{-}catenin) - k_{17r} \cdot (APC/\beta\text{-}catenin) \qquad\qquad 7.124$$

$$v_{18} = vmax_{18} \cdot \frac{1}{K_{HOS}^2 + \left(\dfrac{(\beta\text{-}catenin/TCF)}{k_i}\right)^2} \qquad\qquad 7.125$$

$$v_{19} = k_{19} \cdot (HOS\ mRNA)$$ 7.126

$$v_{20} = k_{20} \cdot (HOS\ mRNA)$$ 7.127

$$v_{21} = k_{21} \cdot (HOS)$$ 7.128

$$v_{22} = k_{22} \cdot (HOS) \cdot (APC^*/Axin^*/GSK3/\beta\text{-}catenin^*)$$ 7.129

$$v_{22b} = k_{22b} \cdot (FWD1) \cdot (APC^*/Axin^*/GSK3/\beta\text{-}catenin^*)$$ 7.130

$$v_{23} = vmax_{23} \cdot \frac{(\beta\text{-}catenin/TCF)^2}{K_{CRD\text{-}BP}^2 + (\beta\text{-}catenin/TCF)^2} + v_{basal}$$ 7.131

$$v_{24} = k_{24} \cdot (CRD\text{-}BP\ mRNA)$$ 7.132

$$v_{25} = k_{25} \cdot (CRD\text{-}BP\ mRNA)$$ 7.133

$$v_{26} = k_{26} \cdot (CRD\text{-}BP)$$ 7.134

$$v_{27} = \frac{k_{27}}{K_{FWD1}^2 + \left(\dfrac{(CRD\text{-}BP)}{k_{i2}}\right)^2} \cdot (FWD1\ mRNA)$$ 7.135

$$v_{28} = constant$$ 7.136

$$v_{29} = k_{29} \cdot (FWD1\ mRNA)$$ 7.137

$$v_{30} = k_{30} \cdot (FWD1)$$ 7.138

$$v_{50} = constant$$ 7.139

$$v_{51} = k_{51} \cdot (TCF)$$ 7.140

7.3.4 Stimulation by Wnt

Absence of Wnt (unstimulated): $Wnt = 0$ 7.141

Constant Wnt stimulation: $Wnt_{t=0} = 0$ and $Wnt_{t>0} = 1$ 7.142

Transient Wnt stimulation: $Wnt_{t=0} = 0$ and $Wnt_{t>0} = Exp[-\lambda \cdot t]$ 7.143

Table 7.6: Parameter of the transient Wnt stimulation.

Parameter	Value	
λ	0.05	min^{-1}

Note, that the Equations 7.141 - 7.143 are identical to those used in the detailed model of Wnt/β-catenin signalling (Appendix 7.2.5, Equations 7.81 - 7.83, respectively).

7.3.5 Model parameters

Table 7.7: Kinetic parameters of the two-feedback model.

Parameter	Value	
k_1	0.182	min^{-1}
k_2	$1.82 \cdot 10^{-2}$	min^{-1}
k_3	$5 \cdot 10^{-2}$	$\text{nM}^{-1} \cdot \text{min}^{-1}$
k_4	0.267	min^{-1}
k_5	0.133	min^{-1}
k_6	$9.09 \cdot 10^{-2}$	$\text{nM}^{-1} \cdot \text{min}^{-1}$
k_{6r}	0.909	min^{-1}
k_7	1	$\text{nM}^{-1} \cdot \text{min}^{-1}$
k_{7r}	50	min^{-1}
k_8	$2.06 \cdot 10^{2}$	$\text{nM}^{-1} \cdot \text{min}^{-1}$
k_{8r}	$2.472 \cdot 10^{4}$	min^{-1}
k_9	$2.06 \cdot 10^{2}$	min^{-1}
k_{9r}	0	min^{-1}
k_{10}	$2.06 \cdot 10^{2}$	min^{-1}
k_{10b}	$2.06 \cdot 10^{2}$	min^{-1}

Table 7.7 continues at the next page.

Table 7.7: Kinetic parameters of the two-feedback model. (continued)

Parameter	Value	
k_{11}	0.417	min^{-1}
v_{12}	0.423	nM·min^{-1}
k_{13}	$2.57·10^{-4}$	min^{-1}
v_{14}	$8.22·10^{-5}$	nM·min^{-1}
k_{15}	0.167	min^{-1}
k_{16}	1	$\text{nM}^{-1}·\text{min}^{-1}$
k_{16r}	30	min^{-1}
k_{17}	1	$\text{nM}^{-1}·\text{min}^{-1}$
k_{17r}	$1.2·10^{3}$	min^{-1}
vmax_{18}	$1·10^{-6}$	nM·min^{-1}
K_{HOS}	1	
k_{19}	$6·10^{-4}$	min^{-1}
k_{20}	30	min^{-1}
k_{21}	$4·10^{-2}$	min^{-1}
k_{22}	$2.06·10^{2}$	$\text{nM}^{-1}·\text{min}^{-1}$
k_{22b}	$2.06·10^{2}$	$\text{nM}^{-1}·\text{min}^{-1}$
vmax_{23}	$6·10^{-6}$	nM·min^{-1}
K_{CRD-BP}	$1.77·10^{2}$	nM
v_{basal}	$1.5·10^{-7}$	nM·min^{-1}
k_{24}	$8.9·10^{-3}$	min^{-1}

Table 7.7 continues at the next page.

Table 7.7: Kinetic parameters of the two-feedback model. (continued)

Parameter	Value	
k_{25}	30	min^{-1}
k_{26}	$2 \cdot 10^{-4}$	min^{-1}
k_{27}	$1 \cdot 10^{-6}$	min^{-1}
K_{FWD1}	$1 \cdot 10^{-2}$	
v_{28}	$1 \cdot 10^{-6}$	$nM \cdot min^{-1}$
k_{29}	30	min^{-1}
k_{30}	$5 \cdot 10^{-3}$	min^{-1}
v_{50}	$2.29 \cdot 10^{-2}$	$nM \cdot min^{-1}$
k_{51}	$2.8 \cdot 10^{-3}$	min^{-1}

Note that the inhibition constants k_i and k_{i2} obtain different values in the analyses of the two-feedback model in Chapter 4. Note furthermore that in most cases, the inverse of these inhibition constants, defined as feedback strength (Section 4.1.3), are considered in that chapter.

Table 7.8: Total protein concentrations of proteins obeying a conservation relation.

Component	Value	
APC_{total}	100	nM
GSK_{total}	50	nM
Dsh_{total}	100	nM

Table 7.9: Kinetic parameters to model mutations of APC.

The parameter values for APC mutants "m1" to "m13" are adopted form (Cho et al., 2006).

APC mutant	k_{7r} [min^{-1}]	k_{8r} [min^{-1}]	k_{17r} [min^{-1}]
"m1"	171	120	1200
"m2"	171	179	1200
"m3"	171	268	1200
"m4"	171	401	1200
"m5"	585	401	1200
"m6"	585	599	1200
"m7"	2000	599	1200
"m8"	2000	895	1200
"m9"	2000	1338	1200
"m10"	2000	2000	1200
"m11"	2000	2000	1423
"m12"	2000	2000	1687
"m13"	2000	2000	2000

7.4 The minimal model of competitive β-TrCP sequestration

The schematic representation of the minimal model is shown in Figure 5.1. In the scheme, the number next to arrow denotes the number of the particular reaction in the model. The model is given by the Equations 7.144 - 7.172. Components in a complex are separated by a slash. Destruction complex is abbreviated with DC as in Figure 5.1. The minimal model is parameterised (see Section 5.2) such that it reproduces the simulations of the detailed models of canonical NF-κB and Wnt/β-catenin signalling (see Section 2.1 and Section 2.2, respectively). In the fitting procedure (see Section 5.2), 400 distinct parameter sets were selected. One randomly chosen parameter set of those 400 is defined to be the reference parameter set and is given in Table 7.10 and Table 7.11.

7.4.1 Differential equations

$$\frac{d(\beta\text{-}catenin)}{dt} = -v_3 + v_5 - v_6 \tag{7.144}$$

$$\frac{d(DC)}{dt} = v_1 - v_2 - v_{2b} - v_3 + v_8 \tag{7.145}$$

$$\frac{d(DC/\beta\text{-}catenin)}{dt} = v_3 - v_7 \tag{7.146}$$

$$\frac{d(DC/\beta\text{-}catenin/\beta\text{-}TrCP)}{dt} = v_7 - v_8 \tag{7.147}$$

$$\frac{d(\beta\text{-}TrCP)}{dt} = -v_7 + v_8 + v_9 - v_{10} - v_{17} + v_{18} \tag{7.148}$$

$$\frac{d(NF\text{-}\kappa B)}{dt} = -v_{13} + v_{18} \tag{7.149}$$

$$\frac{d(I\kappa B)}{dt} = -v_{13} + v_{15} - v_{16} \tag{7.150}$$

$$\frac{d(I\kappa B\ mRNA)}{dt} = v_{11} - v_{12} \tag{7.151}$$

$$\frac{d(NF\text{-}\kappa B/I\kappa B/\beta\text{-}TrCP)}{dt} = v_{17} - v_{18} \tag{7.152}$$

7.4.2 Conservation relation

$$(NF\text{-}\kappa B)_{total} = (NF\text{-}\kappa B) + (NF\text{-}\kappa B/I\kappa B) + (NF\text{-}\kappa B/I\kappa B/\beta\text{-}TrCP) \qquad 7.153$$

7.4.3 Rate equations

$$v_1 = constant \qquad 7.154$$

$$v_2 = k_2 \cdot (DC) \qquad 7.155$$

$$v_{2b} = k_{2b} \cdot Wnt \cdot (DC) \qquad 7.156$$

$$v_3 = k_3 \cdot (DC) \cdot (\beta\text{-}catenin) - k_{3r} \cdot (DC/\beta\text{-}catenin) \qquad 7.157$$

$$v_5 = constant \qquad 7.158$$

$$v_6 = k_6 \cdot (\beta\text{-}catenin) \qquad 7.159$$

$$v_7 = k_7 \cdot (DC/\beta\text{-}catenin) \cdot (\beta\text{-}TrCP) - k_{7r} \cdot (DC/\beta\text{-}catenin/\beta\text{-}TrCP) \qquad 7.160$$

$$v_8 = k_8 \cdot (DC/\beta\text{-}catenin/\beta\text{-}TrCP) \qquad 7.161$$

$$v_9 = constant \qquad 7.162$$

$$v_{10} = k_{10} \cdot (\beta\text{-}TrCP) \qquad 7.163$$

$$v_{11} = vmax_{11} \cdot \frac{(NF\text{-}\kappa B)^2}{K_{11}{}^2 + (NF\text{-}\kappa B)^2} \qquad 7.164$$

$$v_{12} = k_{12} \cdot (I\kappa B\ mRNA) \qquad 7.165$$

$$v_{13} = k_{13} \cdot (NF\text{-}\kappa B) \cdot (I\kappa B) - k_{13r} \cdot (NF\text{-}\kappa B/I\kappa B) \qquad 7.166$$

$$v_{15} = vmax_{15} \cdot \frac{(I\kappa B\ mRNA)^2}{K_{15}{}^2 + (I\kappa B\ mRNA)^2} \qquad 7.167$$

$$v_{16} = k_{16} \cdot (I\kappa B) \qquad 7.168$$

$$v_{17} = k_{17} \cdot TNF \cdot (NF\text{-}\kappa B/I\kappa B) \cdot (\beta\text{-}TrCP) - k_{17r} \cdot (NF\text{-}\kappa B/I\kappa B/\beta\text{-}TrCP) \qquad 7.169$$

$$v_{18} = k_{18} \cdot (NF\text{-}\kappa B/I\kappa B/\beta\text{-}TrCP) \qquad 7.170$$

7.4.4 Model parameters

One parameter set, which is referred to as reference parameter set, was randomly chosen from the 400 selected parameter sets (see Section 5.2.5). This particular parameter set is given in Table 7.10 and Table 7.11.

Table 7.10: Reference set of the parameters of the minimal model.

Parameter	Value	
v_1	$8.22 \cdot 10^{-5}$	$nM \cdot min^{-1}$
k_2	$8.49 \cdot 10^{-3}$	min^{-1}
k_{2b}	$2.30 \cdot 10^{-2}$	min^{-1}
k_3	$1.89 \cdot 10^{4}$	$nM^{-1} \cdot min^{-1}$
k_{3r}	$2.27 \cdot 10^{6}$	min^{-1}
v_5	0.423	$nM \cdot min^{-1}$
k_6	$2.57 \cdot 10^{-4}$	min^{-1}
k_7	$1.32 \cdot 10^{5}$	$nM^{-1} \cdot min^{-1}$
k_{7r}	$1.39 \cdot 10^{2}$	min^{-1}
k_8	$9.47 \cdot 10^{6}$	min^{-1}
v_9	$7.06 \cdot 10^{-8}$	$nM \cdot min^{-1}$
k_{10}	$4.51 \cdot 10^{-5}$	min^{-1}
$vmax_{11}$	$1.40 \cdot 10^{-2}$	$nM \cdot min^{-1}$
K_{11}	49.7	nM
k_{12}	$2.4 \cdot 10^{-2}$	min^{-1}
k_{13}	$9.43 \cdot 10^{-3}$	$nM^{-1} \cdot min^{-1}$
k_{13r}	$2.79 \cdot 10^{-3}$	min^{-1}

Table 7.10 continues at the next page.

Table 7.10: Reference set of the parameters of the minimal model. (continued)

Parameter	Value	
$vmax_{15}$	12.8	$nM \cdot min^{-1}$
K_{15}	0.186	nM
k_{16}	$6 \cdot 10^{-3}$	min^{-1}
k_{17}	$1.28 \cdot 10^{4}$	$nM^{-1} \cdot min^{-1}$
k_{17r}	$1.32 \cdot 10^{6}$	min^{-1}
k_{18}	$6.29 \cdot 10^{3}$	min^{-1}
$(NF\text{-}\kappa B)_{total}$	60	nM

7.4.5 Equations and parameters regarding Wnt and TNF stimulation

$$Wnt_{t=0} = 0 \text{ and } Wnt_{t>0} = Exp\left[-\frac{t}{\tau_0}\right] \tag{7.171}$$

$$TNF = a \cdot \left(\frac{b_1}{1 + c_1 \cdot Exp\left[\frac{t}{\tau_1}\right]} + 1\right) \cdot \left(\frac{b_2}{1 + c_2 \cdot Exp\left[-\frac{t}{\tau_2}\right]} - 1\right) \tag{7.172}$$

Note that the stimulated steady state value of Wnt and TNF is given by the limit of t approaching infinity (see Section 5.2.2). Then, Equation 7.171 and Equation 7.172 simplify to:

$$\lim_{t \to \infty} Wnt = 0 \tag{7.173}$$

$$\lim_{t \to \infty} TNF = a \cdot (b_2 - 1) \tag{7.174}$$

respectively.

Table 7.11: Parameters related to the stimuli of the minimal model.

Parameter	Value	
A	0.86	
b_1	80.9	
b_2	2.47	
c_1	$4.26 \cdot 10^{-2}$	
c_2	1.47	
τ_0	143	Min
τ_1	4.86	Min
τ_2	1.98	Min

8. Additional analyses of the two-feedback model

In Section 4.3.1, the impact of variation of HOS feedback strength on the unstimulated steady state concentration of the β-catenin/TCF complex is investigated under the condition of a disabled FWD1 feedback (scenario I, page 75). Here, the analysis is extended by taking different combinations of HOS and FWD1 feedback strength into account. In the analysis, the number and stability of the unstimulated steady state (Section 8.1) as well as the period length of limit cycle oscillations if detectable (Section 8.2) are explored for different combinations of HOS and FWD1 feedback strength. Note that the results relating to the analysis of the unstimulated steady states of the two-feedback model can also be obtained in an analysis of its stimulated steady states upon transient Wnt stimulation, since the two-feedback model returns to the unstimulated steady state after transient Wnt stimulation (see Section 4.1.5). Hence, for sake of simplicity, the term steady state is used hereafter in this chapter to collectively address the unstimulated steady state as well as the stimulated steady state upon transient Wnt stimulation.

8.1   Number and stability of steady states

To begin with the analysis, the impact of the HOS feedback strength on the steady state of the β-catenin/TCF complex is investigated under the condition of a disabled FWD1 feedback. To that end, the steady state concentration of the β-catenin/TCF complex and the steady state stability (Section 2.7) is calculated for different HOS feedback strengths and plotted in the bifurcation diagram in Figure 8.1A. In the bifurcation diagram, two regions of single stable steady states (Figure 8.1A, black lines) are identified. These two regions are separated by a region of one unstable steady state (grey dashed line in Figure 8.1A) at lower feedback strength (ranging from about 1.013 nM^{-1} to 1.055 nM^{-1}) and a region in which one stable steady state coexists with two unstable steady states (ranging from about 1.055 nM^{-1} to 8.4 nM^{-1}). At the point where the single stable steady state becomes unstable (approximately

1.013 nM^{-1}), a Hopf bifurcation (HB) is detected (Section 2.7). This Hopf bifurcation gives rise to limit cycle of the concentration of the β-catenin/TCF complex around the unstimulated steady state. The minimal and maximal amplitudes of the oscillations are marked black dotted lines in Figure 8.1A. The oscillations are further analysed in Section 8.2.

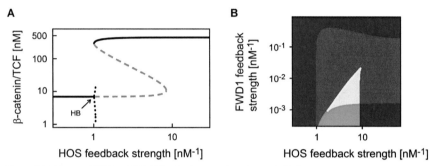

Figure 8.1: Impact of HOS and FWD1 feedback strength on β-catenin/TCF steady state.
(A) The bifurcation diagram characterises the β-catenin/TCF dynamics for different HOS feedback strengths in combination with a disabled FWD1 feedback. Black lines represent stable steady state solutions, grey dashed line represents unstable steady states, and black dotted lines follow the maximal and minimal amplitudes of the stable limit cycles. HB: Hopf bifurcation point. (B) The colour-code of the region plot indicates number and stability of the steady states for different combinations of HOS and FWD1 feedback strength. In the blue areas, single stable β-catenin/TCF steady states are found. In the green area, one stable steady state coexists with two unstable ones. In the red and yellow area, a stable limit cycle coexists with either one or three unstable β-catenin/TCF steady states, respectively.

The bifurcation diagram of Figure 8.1A changes if the FWD1 feedback strength is varied in addition to the HOS feedback strength. The number and stability of the steady states for different combinations of HOS and FWD1 feedback strengths are summarised in the region plot in Figure 8.1B. A region of three coexisting steady states can be detected for different combinations of HOS and FWD1 feedback strength (green and yellow region in Figure 8.1B). Variation of the HOS feedback strength in combination with a disabled FWD1 feedback yields a region of three coexisting steady states. In contrast, if the HOS feedback is disabled and the FWD1 feedback is varied no region of multiple steady states can be found. This indicates that the HOS feedback plays an important role in establishing a region of three coexisting steady states. The analysis in this region of three coexisting steady states reveals that the FWD1 feedback strength influences their stability. For FWD1 feedback strengths below approximately $2 \cdot 10^{-3}$ nM^{-1}, the highest of the three β-catenin/TCF steady state concentrations corresponds to a stable steady state solution, while the other two steady states are instable (Figure 8.1B, green area). In case of FWD1 feedback strengths above approximately $2 \cdot 10^{-3}$ nM^{-1}, all three steady states are unstable (Figure 8.1B, yellow area).

In addition to HOS and FWD1 feedback strength, also other model parameters influence the occurrence of multiple steady states. The following investigation of their impact is by no means a thorough analysis, but rather a pointwise exploration. Despite that, the results are mentioned here as they offer further mechanistic explanations of the dynamical behaviour of the two-feedback model. The first result is obtained from a sensitivity analysis of each model parameter on the steady state concentration of the β-catenin/TCF complex for different combinations of HOS and FWD1 feedback strengths. The analysis indicates that the sensitivity coefficients (Section 2.6) of the parameters involved in the HOS feedback ($vmax_{18}$, k_{19}, k_{20}, and k_{21}) can be reduced by increasing the FWD1 feedback strength (not shown). The reduction of the sensitivity coefficients is paralleled by a decrease in the range of the HOS feedback strengths that yields three coexisting steady states. The region of three coexisting steady states vanishes at the FWD1 feedback strengths of approximately $2.1 \cdot 10^{-2}$ nM^{-1} (Figure 8.1B, yellow area). The combination of HOS and FWD1 feedback strengths of approximately 9.1 nM^{-1} and $2.1 \cdot 10^{-2}$ nM^{-1}, respectively, may thus indicate the occurrence of a cusp catastrophe. Similar to increasing FWD1 feedback strength, an increased rate constant of alternative β-catenin degradation k_{13} (Reaction 13 in Figure 4.1) reduces the sensitivity coefficients of the parameters related to the HOS feedback on the steady state of the β-catenin/TCF complex (not shown). Simulations revealed that already an increase of rate constant of alternative β-catenin degradation by less than two fold abolishes the existence of multiple steady state solutions for all tested combinations of HOS and FWD1 feedback strength (not shown).

The existence of multiple steady state solutions also depends on the conservation relations of GSK3 and APC (Equation 7.105 and Equation 7.106, respectively, in Section 7.3.2). If either the APC or the GSK3 conservation relation is replaced by a production and degradation reaction of APC or GSK3, respectively, three coexisting steady states are found for certain combinations of HOS and FWD1 feedback strength. However, that is not the case if both conservation relations are simultaneously replaced (not shown).

Taken together, the choice of a particular combination of HOS and FWD1 feedback strength influences the number and stability of the steady states. In addition, parameters that reduce the sensitivity coefficients of the parameters related to the HOS feedback on the steady state of the β-catenin/TCF complex may also reduce the range of HOS feedback strengths that yields three coexisting steady states.

8.2   Analysis of the limit cycle oscillations

In the case of disabled FWD1 feedback and increasing HOS feedback strength, the stable steady state of the β-catenin/TCF complex becomes instable passing a Hopf bifurcation point at HOS feedback strength of about 1.013 nM^{-1} (Figure 8.1A). At that Hopf bifurcation point, limit cycle oscillations emerge. The amplitude of the oscillations increase rapidly with increasing HOS feedback strength (Figure 8.1A, black dotted lines). Similarly, the period length of the oscillations increases strongly with increasing HOS feedback strength (Figure 8.2A). The minimal period length of about 110 h already exceeds the signal duration of Wnt/β-catenin signalling (Figure 4.2) by about 20-fold. The oscillations therefore seem irrelevant in the time range of Wnt/β-catenin signalling. If the increasing HOS feedback strength approaches the region of three coexisting steady states (Figure 8.1A), the period lengths increase exponentially (Figure 8.2A), until the limit cycles finally vanish at a HOS feedback strength of about 1.055 nM^{-1} (Figure 8.1A).

Simulations reveal that the limit cycle oscillations are relaxation oscillations. They are characterised by a long time range in the order of days with high concentrations of the β-catenin/TCF complex that alternates with a short time range in the order of minutes with low concentrations of the β-catenin/TCF complex. The actual minimal and maximal concentration depends on the HOS feedback strength (Figure 8.1A, black dotted lines). One possible mechanism how relaxation oscillations arise may be a transient interruption of β-catenin degradation by the destruction complex. This mechanism could work as follows: accumulation of the β-catenin/TCF complex results in the decrease of the HOS concentration that blocks the destruction cycle. Subsequently, the β-catenin*/APC*/Axin*/GSK3 complex accumulates. If the concentration of this complex is high enough, it allows for the degradation of β-catenin via the destruction core cycle even in the presence of low amounts of HOS. This accumulation process of the β-catenin*/APC*/Axin*/GSK3 complex counteracting the down-regulation of HOS may be one possible mechanistic explanation for the oscillations. To test this hypothesis, the irreversible phosphorylation of β-catenin in the destruction complex (Reaction 9 in Figure 4.1) is changed to be reversible by setting the parameter k_{9r}

greater zero (Equation 7.115, Appendix 7.3). Consequently, the accumulation of the β-catenin*/APC*/Axin*/GSK3 complex is impaired. Simulations reveal that increasing values of parameter k_{9r} narrow the region of HOS feedback strengths in which limit cycle oscillations occur. In case of a parameter value of approximately 2 min^{-1}, no Hopf bifurcation is found and the response of the β-catenin/TCF steady state on the HOS feedback strength becomes bistable (not shown). The results indicate that transient interruption of β-catenin degradation may be a possible mechanism how relaxation oscillations arise.

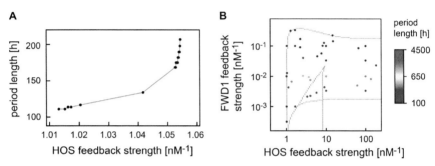

Figure 8.2: Impact of the feedback strengths on the period length of the limit cycle oscillations.
(A) The period length of the limit cycles is plotted for different HOS feedback strengths. The FWD1 feedback is disabled in all simulations. (B) For different combinations of HOS and FWD1 feedback strengths, the period length of the limit cycle oscillation was estimated. The calculated period lengths are indicated by colour-coded dots. A legend of the colour-code is provided on the right hand side. The dotted grey lines mark the borders of the distinct areas described in Figure 8.1B.

Next, the period length of the limit cycle oscillations for different combinations of HOS and FWD1 feedback strength is investigated (Figure 8.2B). The period length of the limit cycle oscillations shortens by up to one order of magnitude for increasing FWD1 feedback strengths at a fixed HOS feedback strength. This may be explained as follows. A HOS feedback strength larger than about 1 nM^{-1} (Figure 8.2B) interrupts β-catenin degradation via the β-catenin*/APC*/Axin*/GSK3/HOS complex by reducing HOS expression. This leads to further increase of β-catenin/TCF complex concentration. If the FWD1 feedback is not disabled, rising β-catenin/TCF complex concentrations increase FWD1 concentrations. Due to the increase of FWD1 concentrations, β-catenin can be decreased via the β-catenin*/APC*/Axin*/GSK3/FWD1 complex despite low concentrations of HOS. Consequently, the period length shortens with increasing FWD1 feedback strength (Figure 8.2B). The period length at the Hopf bifurcation point (HOS feedback strength of approximately 1 nM^{-1}), however, seems to be independent of the FWD1 feedback strength as indicated by the blue-coloured dots at FWD1 feedback strengths of 10^{-3} nM^{-1}, 10^{-2} nM^{-1}, and 10^{-1} nM^{-1} in Figure 8.2B.

9. Additional analyses of the minimal model of competitive β-TrCP sequestration

9.1 Steady state analysis of the minimal model of competitive β-TrCP sequestration

The steady state of a species is given by the exact counterbalance of all producing and converting processes associated with this species. This is equivalent to setting the temporal derivative of the concentration of each species to zero in ODE models. Taking the ODE system of the minimal model into account, the ODE system (Equations 7.144 - 7.152) yield the following new set of algebraic equations (Equations 9.1 - 9.9) describing the steady state of the model.

$$0 = -v_3 + v_5 - v_6 \tag{9.1}$$

$$0 = v_1 - v_2 - v_{2b} - v_3 + v_8 \tag{9.2}$$

$$0 = v_3 - v_7 \tag{9.3}$$

$$0 = v_7 - v_8 \tag{9.4}$$

$$0 = -v_7 + v_8 + v_9 - v_{10} - v_{17} + v_{18} \tag{9.5}$$

$$0 = -v_{13} + v_{18} \tag{9.6}$$

$$0 = -v_{13} + v_{15} - v_{16} \tag{9.7}$$

$$0 = v_{11} - v_{12} \tag{9.8}$$

$$0 = v_{17} - v_{18} \tag{9.9}$$

Furthermore, the conservation relation of NF-κB (Equation 7.153) has to be fulfilled. In the following algebraic transformations, the dependences of the steady states on the parameters are derived.

Considering Equation 9.4 and Equation 9.9, Equation 9.5 simplifies to

$$0 = v_9 - v_{10}$$
9.10

By substituting v_{10} in Equation 9.10 by Equation 7.163, the steady state solution of β-TrCP is derived (Equation 9.11):

$$(\beta\text{-}TrCP)_{stst} = \frac{v_9}{k_{10}}$$
9.11

The steady state solution of the destruction complex (DC) (Equation 9.12) is obtained by adding Equation 9.2, Equation 9.3, and Equation 9.4 and subsequently substituting v_2 and v_{2b} by Equation 7.155 and Equation 7.156, respectively:

$$(DC)_{stst} = \frac{v_1}{k_2 + k_{2b} \cdot Wnt}$$
9.12

Equation 7.171 and Equation 7.173 show that the Wnt stimulus in case of the unstimulated (i.e., $t = 0$) and the transiently stimulated (i.e., $t \to \infty$) steady state of the model equals zero. Thus Equation 9.12 further simplifies to:

$$(DC)_{stst} = \frac{v_1}{k_2}$$
9.13

Equation 9.3 is simplified by substituting the rates v_3 and v_7 by Equation 7.157 and Equation 7.160, respectively, and taking the steady state Equation 9.11 and Equation 9.13 into account to:

$$(DC/\beta\text{-}catenin)_{stst} = \frac{k_3 \cdot \frac{v_1}{k_2} \cdot (\beta\text{-}catenin)_{stst} + k_{7r} \cdot (DC/\beta\text{-}catenin/\beta\text{-}TrCP)_{stst}}{k_{3r} + k_7 \cdot \frac{v_9}{k_{10}}}$$
9.14

Equation 9.4 is simplified by substituting the rates v_7 and v_8 by Equation 7.160 and Equation 7.161, respectively, and taking the steady state Equation 9.11 and Equation 9.13 into account to:

$$(DC/\beta\text{-}catenin)_{stst} = \frac{(k_{7r} + k_8)}{k_7} \cdot \frac{(DC/\beta\text{-}catenin/\beta\text{-}TrCP)_{stst}}{\frac{v_9}{k_{10}}}$$

9.15

The Equation 9.16 of the β-catenin steady state is derived by adding Equation 9.1, Equation 9.3, and Equation 9.4 and subsequently substituting v_6 and v_8 by Equation 7.159 and Equation 7.161, respectively.

$$(\beta\text{-}catenin)_{stst} = \frac{v_5 - k_8 \cdot (destruction\ complex/\beta\text{-}catenin/\beta\text{-}TrCP)_{stst}}{k_6}$$

9.16

Solving Equations 9.14 - 9.16 yields the steady state solutions of β-catenin, DC/β-catenin, and DC/β-catenin/β-TrCP (Equations 9.17 - 9.19):

$$(\beta\text{-}catenin)_{stst} = \frac{v_5}{k_6} \cdot \frac{\frac{k_{3r}}{k_3} + \frac{k_7}{k_{7r} + k_8} \cdot \frac{k_8}{k_3} \cdot \frac{v_9}{k_{10}}}{\frac{k_{3r}}{k_3} + \left(\frac{v_1}{k_2} \cdot \frac{k_3}{k_6} + 1\right) \cdot \frac{k_7}{k_{7r} + k_8} \cdot \frac{k_8}{k_3} \cdot \frac{v_9}{k_{10}}}$$

9.17

$$(DC/\beta\text{-}catenin)_{stst} = \frac{\frac{v_1}{k_2} \cdot \frac{v_5}{k_6}}{\frac{k_{3r}}{k_3} + \left(\frac{v_1}{k_2} \cdot \frac{k_3}{k_6} + 1\right) \cdot \frac{k_7}{k_{7r} + k_8} \cdot \frac{k_8}{k_3} \cdot \frac{v_9}{k_{10}}}$$

9.18

$$(DC/\beta\text{-}catenin/\beta\text{-}TrCP)_{stst} = \frac{\frac{v_1}{k_2} \cdot \frac{v_5}{k_6} \cdot \frac{k_7}{k_{7r} + k_8} \cdot \frac{v_9}{k_{10}}}{\frac{k_{3r}}{k_3} + \left(\frac{v_1}{k_2} \cdot \frac{k_3}{k_6} + 1\right) \cdot \frac{k_7}{k_{7r} + k_8} \cdot \frac{k_8}{k_3} \cdot \frac{v_9}{k_{10}}}$$

9.19

Equation 9.8 is simplified by substituting the rates v_{11} and v_{12} by Equation 7.164 and Equation 7.165, respectively, to yield the steady state of IκB mRNA:

$$(I\kappa B\ mRNA)_{stst} = \frac{vmax_{11}}{k_{12}} \cdot \frac{((NF\text{-}\kappa B)_{stst})^2}{K_{11}^2 + ((NF\text{-}\kappa B)_{stst})^2}$$

9.20

The simplified Equation 9.21 of the IκB steady state is derived by subtracting Equation 9.6 from Equation 9.7, and subsequently substituting v_{15}, v_{16}, and v_{18} by Equation 7.167, Equation 7.168, and Equation 7.170, respectively.

$$(I\kappa B)_{stst} = \frac{vmax_{15}}{k_{16}} \cdot \frac{((I\kappa B\text{-}mRNA)_{stst})^2}{K_{15}^2 + ((I\kappa B\text{-}mRNA)_{stst})^2} - \frac{k_{18}}{k_{16}} \cdot (NF\text{-}\kappa B/I\kappa B/\beta\text{-}TrCP)_{stst} \qquad 9.21$$

Equation 9.9 is simplified by substituting the rates v_{17} and v_{18} by Equation 7.169 and Equation 7.170, respectively, and taking the conservation relation of NF-κB (Equation 7.153) into account, to:

$$(NF\text{-}\kappa B/I\kappa B/\beta\text{-}TrCP)_{stst} = \frac{\dfrac{k_{17}}{k_{17r}+k_{18}} \cdot \dfrac{v_9}{k_{10}} \cdot ((NF\text{-}\kappa B)_{total} - (NF\text{-}\kappa B)_{stst}) \cdot TNF_{stst}}{1 + \dfrac{k_{17}}{k_{17r}+k_{18}} \cdot \dfrac{v_9}{k_{10}} \cdot TNF_{stst}} \qquad 9.22$$

Equation 9.6 is simplified by substituting the rates v_{13} and v_{18} by Equation 7.166 and Equation 7.170, respectively, and taking into account the conservation relation of NF-κB (Equation 7.153) as well as Equations 9.20 - 9.22, to yield the following polynomial in $(NF\text{-}\kappa B)_{stst}$:

$$0 = \frac{k_{13r}}{k_{13}} \cdot ((NF\text{-}\kappa B)_{total} - (NF\text{-}\kappa B)_{stst})$$

$$+ \left(\frac{k_{18}}{k_{13}} - \frac{k_{13r}}{k_{13}}\right) \cdot \frac{\dfrac{k_{17}}{k_{17r}+k_{18}} \cdot \dfrac{v_9}{k_{10}} \cdot TNF_{stst}}{1 + \dfrac{k_{17}}{k_{17r}+k_{18}} \cdot \dfrac{v_9}{k_{10}} \cdot TNF_{stst}} \cdot ((NF\text{-}\kappa B)_{total} - (NF\text{-}\kappa B)_{stst})$$

$$- \frac{vmax_{15}}{k_{16}} \cdot \frac{\dfrac{((NF\text{-}\kappa B)_{stst})^4}{(K_{11}^2 + ((NF\text{-}\kappa B)_{stst})^2)^2}}{\left(\dfrac{k_{12} \cdot K_{15}}{vmax_{11}}\right)^2 + \dfrac{((NF\text{-}\kappa B)_{stst})^4}{(K_{11}^2 + ((NF\text{-}\kappa B)_{stst})^2)^2}} \cdot (NF\text{-}\kappa B)_{stst}$$

$$+ \frac{k_{18}}{k_{16}} \cdot \frac{\dfrac{k_{17}}{k_{17r}+k_{18}} \cdot \dfrac{v_9}{k_{10}} \cdot TNF_{stst}}{1 + \dfrac{k_{17}}{k_{17r}+k_{18}} \cdot \dfrac{v_9}{k_{10}} \cdot TNF_{stst}} \cdot ((NF\text{-}\kappa B)_{total} - (NF\text{-}\kappa B)_{stst}) \cdot (NF\text{-}\kappa B)_{stst} \qquad 9.23$$

9.2 Derivation of parameter sets rescaled in time

In Section 5.5.2, the effects of variations in relative signalling time of the canonical NF-κB signalling module and the Wnt/β-catenin signalling module on the direction of crosstalk in the minimal model are investigated for the reference parameter set (Table 7.10 and Table 7.11, Appendix 7.4). Three different cases of relative timing between the two signalling modules are considered: (i) the signalling time of β-catenin dynamics is larger than that of NF-κB dynamics (reference situation: 285.9 min and 75.0 min, respectively; Table 5.5), (ii) the signalling time of β-catenin and NF-κB dynamics are similar (84.1 min and 85.3 min, respectively), and (iii) the signalling time of β-catenin dynamics is smaller than that of NF-κB dynamics (75.0 min and 285.5 min , respectively). The reference situation (i) is obtained in simulations of the minimal model using the reference parameter set (Table 7.10 and Table 7.11, Appendix 7.4). To obtain cases (ii) and (iii) with different signalling times of the two pathway modules, the reference parameter set is changed. To that end, the rate equations v_1 to v_8 of the Wnt/β-catenin signalling module (Equations 7.154 - 7.161, Appendix 7.4) are multiplied by a scaling factor of 3.4 in case (ii) and 3.81 in case (iii). In addition, the parameter τ_0 of the transient Wnt stimulus (Equation 7.171, Appendix 7.4) is divided by these scaling factors (3.4 and 3.81 in case (ii) and (iii), respectively). Considering the canonical NF-κB pathway module, the rate equations v_{11} to v_{18} (Equations 7.164 - 7.170, Appendix 7.4) are multiplied by a scaling factor of 0.8 in case (ii) and 0.162 in case (iii). In addition, the parameters τ_1 and τ_2 of the transient TNF stimulus (Equation 7.172, Appendix 7.4) are divided by these scaling factors (0.8 and 0.162 in case (ii) and (iii), respectively). The production and degradation rates of β-TrCP (v_9 and v_{10}, respectively) remain unchanged. The applied scaling factors were derived by manual fitting to obtain the desired signalling time in the simulations.

9.3 Derivation of the parameter set rescaled in concentration

In Section 5.5.3, the effects on the direction of crosstalk are investigated if the concentrations of the pathway components of the canonical NF-κB and the Wnt/β-catenin signalling module are adapted to each other by rescaling model parameters. As in Section 5.5.2, again the reference parameter set (Table 7.10 and Table 7.11, Appendix 7.4) constitutes the starting point of the parameter manipulations in this investigation.

First, the concentrations of β-TrCP and the species of the canonical NF-κB pathway module (Figure 5.1, black and red, respectively) are scaled in concentration by a factor of $2 \cdot 10^{-6}$. This factor relates to the volume ratio of the two experimental model systems for which the detailed models of Wnt/β-catenin and canonical NF-κB signalling have been established. That is, xenopus oocytes with an average volume of approximately 1 μL (Sims and Allbritton, 2007) and mouse embryonic fibroblasts (MEFs) with an average volume of about 2 pL (de Groof et al., 2009; Lang et al., 1992), respectively. This parameter scaling approach can thus be interpreted as normalisation of the concentrations of the species in the minimal model on the same cellular volume. To obtain the scaled concentrations of β-TrCP and the species of the canonical NF-κB pathway module, the parameters v_9, $vmax_{11}$, $vmax_{15}$, K_{15}, K_{11}, and $(NF\text{-}\kappa B)_{total}$ are multiplied by the scaling factor of $2 \cdot 10^{-6}$ and the parameters k_7, k_{13}, and k_{17} are divided by the factor of $2 \cdot 10^{-6}$. These changes in the parameters result in changes of the β-catenin dynamics upon transient Wnt stimulation. In particular, the signal amplitude decreases to 6.0 nM. In addition, signalling time and signal duration increase to 800 min and 870 min, respectively. To compensate this change in the β-catenin dynamics upon transient Wnt stimulation, the rate equations v_1 to v_8 of the Wnt/β-catenin signalling module (Equations 7.154 - 7.161, Appendix 7.4) are multiplied by a factor of 0.15. In addition, the parameter τ_0 of the transient Wnt stimulus (Equation 7.171, Appendix 7.4) is divided by a factor of 0.6. These two scaling factors (0.15 and 0.6) were derived by manual fitting to obtain β-catenin dynamics upon transient Wnt stimulation (Figure 5.13A, grey line) similar to the simulations for the reference parameter set (Figure 5.11A, grey line). For the resulting rescaled parameter set, a signal amplitude of 21.6 nM, a signalling time of 329.5 min, and a signal duration of 359.3 min are calculated, which is close to the values obtained for the reference parameter set (Table 5.5, Figure 5.11A).

9.4 Tables listing Kendall rank correlation coefficients analysed in Chapter 5

Table 9.1: Kendall rank correlation coefficients of Figure 5.6A and their corresponding p-values.

Kendall rank correlation coefficients (Section 2.8) between parameter combinations and minimal stimulated steady state of NF-κB in the case of β-TrCP approaching zero are listed. The Kendall rank correlation coefficients of the indicated parameter combinations are sorted with respect to their corresponding p-values.

parameter combinations	Kendall rank correlation coefficient	p-value
$\dfrac{k_{13r}}{k_{13}}$	0.95	$1.7 \cdot 10^{-178}$
K_{11}	-0.17	$2.2 \cdot 10^{-7}$
$K_{15} \cdot \dfrac{k_{12}}{vmax_{11}}$	0.17	$4.1 \cdot 10^{-7}$
$\dfrac{vmax_{15}}{k_{16}}$	0.17	$6.5 \cdot 10^{-7}$
$\dfrac{k_{18}}{k_{16}}$	0.04	0.20
$\dfrac{k_{18}}{k_{13}}$	0.04	0.22

Table 9.2: Kendall rank correlation coefficients of Figure 5.6B and their corresponding p-values.

Kendall rank correlation coefficients (Section 2.8) between parameter combinations and maximal stimulated steady state of NF-κB in the case of β-TrCP approaching infinity are listed. The Kendall rank correlation coefficients of the indicated parameter combinations are sorted with respect to their corresponding p-values.

parameter combinations	Kendall rank correlation coefficient	p-value
$\dfrac{k_{18}}{k_{16}}$	0.97	$6.3 \cdot 10^{-183}$
$\dfrac{k_{18}}{k_{13}}$	0.95	$7.5 \cdot 10^{-179}$
K_{11}	-0.22	$8.2 \cdot 10^{-11}$
$K_{15} \cdot \dfrac{k_{12}}{vmax_{11}}$	0.11	$9.1 \cdot 10^{-4}$
$\dfrac{vmax_{15}}{k_{16}}$	0.05	0.13
$\dfrac{k_{13r}}{k_{13}}$	0.05	0.14

Table 9.3: Kendall rank correlation coefficients and their corresponding p-values (Section 5.3.5).

Kendall rank correlation coefficients and corresponding p-values (Section 2.8) are listed for the parameters and parameter combinations vs. the respective CIs (Section 2.5) for all 400 selected parameter sets. The Kendall rank correlation coefficients are sorted with respect to their corresponding p-values.

parameter/parameter combination	Kendall rank correlation coefficient	p-value
v_9	-0.53	$1.3 \cdot 10^{-53}$
$\dfrac{k_{17}}{k_{17r} + k_{18}}$	0.48	$6.5 \cdot 10^{-44}$
k_{17}	0.36	$4.2 \cdot 10^{-26}$
k_7	0.36	$1.5 \cdot 10^{-25}$
$\dfrac{v_9}{k_{10}}$	-0.35	$3.7 \cdot 10^{-25}$
$\dfrac{k_7}{k_{7r} + k_8}$	0.35	$6.2 \cdot 10^{-24}$
k_{10}	-0.22	$1.6 \cdot 10^{-10}$
$vmax_{11}$	0.13	$1.1 \cdot 10^{-4}$
$\dfrac{k_{18}}{k_{16}}$	-0.12	$3.8 \cdot 10^{-4}$
k_{18}	-0.12	$3.8 \cdot 10^{-4}$
$\dfrac{k_{18}}{k_{13}}$	-0.12	$4.3 \cdot 10^{-4}$
K_{11}	0.12	$5.9 \cdot 10^{-4}$
$\dfrac{k_{3r}}{k_3}$	0.11	$5.5 \cdot 10^{-3}$
k_8	-0.06	$8.2 \cdot 10^{-2}$
k_{13r}	-0.06	$9.4 \cdot 10^{-2}$
$\dfrac{k_{13r}}{k_{13}}$	-0.06	$9.5 \cdot 10^{-2}$

Table 9.3 continues at the next page.

Table 9.3: Kendall rank correlation coefficients and their corresponding p-values (Section 5.3.5). (continued)

parameter/parameter combination	Kendall rank correlation coefficient	p-value
$K_{15} \cdot \dfrac{k_{12}}{vmax_{11}}$	-0.05	0.16
k_{13}	-0.04	0.25
$\dfrac{k_8}{k_3}$	-0.04	0.27
λ	-0.03	0.44
k_{2b}	0.02	0.48
$\dfrac{vmax_{15}}{k_{16}}$	-0.02	0.64
$vmax_{15}$	-0.02	0.64
K_{15}	-0.01	0.68
$\dfrac{k_3}{k_6}$	0.01	0.71
k_{3r}	0.01	0.71
k_3	0.01	0.71
k_{7r}	-0.01	0.75
k_{17r}	-0.01	0.81

Table 9.4: Kendall rank correlation coefficients and their corresponding p-values for reduced parameter values of v_9 and k_{10} (Section 5.3.6).

Kendall rank correlation coefficients and corresponding p-values (Section 2.8) are listed for the parameters and parameter combinations vs. the respective CIs (Section 2.5) The reduced β-TrCP production (v_9) and degradation rate constants (k_{10}) are calculated for all 400 selected parameter sets using Equation 5.6 and Equation 5.7, respectively (see Section 5.3.6). The Kendall rank correlation coefficients are sorted with respect to their corresponding p-values.

parameter/parameter combination	Kendall rank correlation coefficient	p-value
$\dfrac{k_{17}}{k_{17r} + k_{18}}$	0.96	$5.0 \cdot 10^{-167}$
v_9	-0.53	$3.8 \cdot 10^{-52}$
$\dfrac{v_9}{k_{10}}$	-0.53	$3.8 \cdot 10^{-52}$
k_{10}	0.53	$3.8 \cdot 10^{-52}$
k_{17}	0.51	$1.1 \cdot 10^{-47}$
k_7	0.49	$6.3 \cdot 10^{-45}$
$\dfrac{k_7}{k_{7r} + k_8}$	0.49	$1.1 \cdot 10^{-44}$
$\dfrac{k_{18}}{k_{16}}$	-0.28	$6.3 \cdot 10^{-16}$
k_{18}	-0.28	$6.3 \cdot 10^{-16}$
$\dfrac{k_{18}}{k_{13}}$	-0.28	$6.3 \cdot 10^{-16}$
$vmax_{11}$	0.23	$1.0 \cdot 10^{-10}$
K_{11}	0.21	$3.4 \cdot 10^{-9}$
$K_{15} \cdot \dfrac{k_{12}}{vmax_{11}}$	-0.14	$9.2 \cdot 10^{-5}$
$\dfrac{k_3}{k_6}$	-0.11	$1.4 \cdot 10^{-3}$
k_{3r}	-0.11	$1.4 \cdot 10^{-3}$

Table 9.4 continues at the next page.

Table 9.4: Kendall rank correlation coefficients and their corresponding p-values for reduced parameter values of v_9 and k_{10} (Section 5.3.6). (continued)

parameter/parameter combination	Kendall rank correlation coefficient	p-value
k_3	-0.11	$1.4 \cdot 10^{-3}$
$\dfrac{vmax_{15}}{k_{16}}$	-0.09	$7.8 \cdot 10^{-3}$
$vmax_{15}$	-0.09	$7.8 \cdot 10^{-3}$
k_{13}	-0.09	$1.0 \cdot 10^{-2}$
K_{15}	-0.07	$3.2 \cdot 10^{-2}$
k_8	-0.07	$5.0 \cdot 10^{-2}$
k_{7r}	-0.06	$7.3 \cdot 10^{-2}$
k_{17r}	-0.05	0.15
$\dfrac{k_{3r}}{k_3}$	0.01	0.71
$\dfrac{k_8}{k_3}$	0.01	0.86
k_{13r}	-0.00	0.90
k_{2b}	-0.00	0.92
$\dfrac{k_{13r}}{k_{13}}$	-0.00	0.93
λ	-0.00	0.96

9.5   Dynamics of β-catenin considering a long time range.

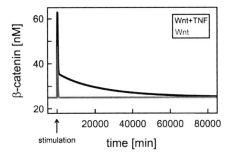

Figure 9.1: Effects of simultaneous TNF and Wnt stimulation on the dynamics of β-catenin considering a long time range.

Dynamics of β-catenin upon Wnt stimulation (grey line) is compared to dynamics of β-catenin upon simultaneous stimulation with Wnt and TNF (black line). In contrast to Figure 5.7A in Section 5.3.4, the transient β-catenin dynamics ranging from initial to final steady state of β-catenin is shown. Simulations consider the reference parameter set (Table 7.10 and Table 7.11, Appendix 7.4.4).

List of Figures

List of Tables

Bibliography

Aberle, H., Bauer, A., Stappert, J., Kispert, A., and Kemler, R. (1997). beta-catenin is a target for the ubiquitin-proteasome pathway. EMBO J *16*, 3797-3804.

Albanese, C., Wu, K., D'Amico, M., Jarrett, C., Joyce, D., Hughes, J., Hulit, J., Sakamaki, T., Fu, M., Ben-Ze'ev, A., *et al.* (2003). IKKalpha regulates mitogenic signaling through transcriptional induction of cyclin D1 via Tcf. Mol Biol Cell *14*, 585-599.

Alkalay, I., Yaron, A., Hatzubai, A., Orian, A., Ciechanover, A., and Ben-Neriah, Y. (1995). Stimulation-dependent I kappa B alpha phosphorylation marks the NF-kappa B inhibitor for degradation via the ubiquitin-proteasome pathway. Proceedings of the National Academy of Sciences *92*, 10599-10603.

Amit, S., and Ben-Neriah, Y. (2003). NF-kappaB activation in cancer: a challenge for ubiquitination- and proteasome-based therapeutic approach. Semin Cancer Biol *13*, 15-28.

Archbold, H.C., Yang, Y.X., Chen, L., and Cadigan, K.M. (2012). How do they do Wnt they do?: regulation of transcription by the Wnt/beta-catenin pathway. Acta Physiol (Oxf) *204*, 74-109.

Aulehla, A., and Herrmann, B.G. (2004). Segmentation in vertebrates: clock and gradient finally joined. Genes & Development *18*, 2060-2067.

Aulehla, A., and Pourquié, O. (2008). Oscillating signaling pathways during embryonic development. Current Opinion in Cell Biology *20*, 632-637.

Bai, C., Sen, P., Hofmann, K., Ma, L., Goebl, M., Harper, J.W., and Elledge, S.J. (1996). SKP1 Connects Cell Cycle Regulators to the Ubiquitin Proteolysis Machinery through a Novel Motif, the F-Box. Cell *86*, 263-274.

Ballarino, M., Fruscalzo, A., Marchioni, M., and Carnevali, F. (2004). Identification of positive and negative regulatory regions controlling expression of the Xenopus laevis betaTrCP gene. Gene *336*, 275-285.

Bardwell, L., Zou, X., Nie, Q., and Komarova, N.L. (2007). Mathematical models of specificity in cell signaling. Biophys J *92*, 3425-3441.

Barken, D., Wang, C.J., Kearns, J., Cheong, R., Hoffmann, A., and Levchenko, A. (2005). Comment on "Oscillations in NF-kappaB Signaling Control the Dynamics of Gene Expression". Science *308*, 52.

Barker, N., and Clevers, H. (2006). Mining the Wnt pathway for cancer therapeutics. Nat Rev Drug Discov *5*, 997-1014.

Basak, S., Behar, M., and Hoffmann, A. (2012). Lessons from mathematically modeling the NF-κB pathway. Immunological Reviews *246*, 221-238.

Behar, M., Dohlman, H.G., and Elston, T.C. (2007). Kinetic insulation as an effective mechanism for achieving pathway specificity in intracellular signaling networks. Proc Natl Acad Sci U S A *104*, 16146-16151.

Belaidouni, N., Peuchmaur, M., Perret, C., Florentin, A., Benarous, R., and Besnard-Guerin, C. (2005). Overexpression of human beta TrCP1 deleted of its F box induces tumorigenesis in transgenic mice. Oncogene *24*, 2271-2276.

Ben-Neriah, Y., and Karin, M. (2011). Inflammation meets cancer, with NF-kappaB as the matchmaker. Nat Immunol *12*, 715-723.

Benary, U., Kofahl, B., Hecht, A., and Wolf, J. (2013). Modelling Wnt/ß-catenin target gene expression in APC and Wnt gradients under wild type and mutant conditions. Frontiers in Physiology *4*.

Besnard-Guerin, C., Belaidouni, N., Lassot, I., Segeral, E., Jobart, A., Marchal, C., and Benarous, R. (2004). HIV-1 Vpu sequesters beta-transducin repeat-containing protein (betaTrCP) in the cytoplasm and provokes the accumulation of beta-catenin and other SCFbetaTrCP substrates. J Biol Chem *279*, 788-795.

Bhatia, N., Herter, J.R., Slaga, T.J., Fuchs, S.Y., and Spiegelman, V.S. (2002). Mouse homologue of HOS (mHOS) is overexpressed in skin tumors and implicated in constitutive activation of NF-kappaB. Oncogene *21*, 1501-1509.

Bianchi, K., and Meier, P. (2009). A tangled web of ubiquitin chains: breaking news in TNF-R1 signaling. Mol Cell *36*, 736-742.

Billadeau, D.D. (2007). Primers on molecular pathways. The glycogen synthase kinase-3beta. Pancreatology *7*, 398-402.

Bour, S., Perrin, C., Akari, H., and Strebel, K. (2001). The human immunodeficiency virus type 1 Vpu protein inhibits NF-kappa B activation by interfering with beta TrCP-mediated degradation of Ikappa B. J Biol Chem *276*, 15920-15928.

Bournat, J.C., Brown, A.M., and Soler, A.P. (2000). Wnt-1 dependent activation of the survival factor NF-kappaB in PC12 cells. J Neurosci Res *61*, 21-32.

Cadigan, K.M., and Waterman, M.L. (2012). TCF/LEFs and Wnt Signaling in the Nucleus. Cold Spring Harb Perspect Biol *4*.

Carayol, N., and Wang, C.Y. (2006). IKKalpha stabilizes cytosolic beta-catenin by inhibiting both canonical and non-canonical degradation pathways. Cell Signal *18*, 1941-1946.

Chechik, G., and Koller, D. (2009). Timing of gene expression responses to environmental changes. J Comput Biol *16*, 279-290.

Chen, Y., Gruidl, M., Remily-Wood, E., Liu, R.Z., Eschrich, S., Lloyd, M., Nasir, A., Bui, M.M., Huang, E., Shibata, D., *et al.* (2010). Quantification of beta-catenin signaling components in colon cancer cell lines, tissue sections, and microdissected tumor cells using reaction monitoring mass spectrometry. J Proteome Res *9*, 4215-4227.

Chen, Z., Hagler, J., Palombella, V.J., Melandri, F., Scherer, D., Ballard, D., and Maniatis, T. (1995). Signal-induced site-specific phosphorylation targets I kappa B alpha to the ubiquitin-proteasome pathway. Genes & Development *9*, 1586-1597.

Chen, Z.J. (2012). Ubiquitination in signaling to and activation of IKK. Immunological Reviews *246*, 95-106.

Cheong, R., Bergmann, A., Werner, S.L., Regal, J., Hoffmann, A., and Levchenko, A. (2006). Transient IkappaB kinase activity mediates temporal NF-kappaB dynamics in response to a wide range of tumor necrosis factor-alpha doses. J Biol Chem *281*, 2945-2950.

Cheong, R., Hoffmann, A., and Levchenko, A. (2008). Understanding NF-kappaB signaling via mathematical modeling. Mol Syst Biol *4*, 192.

Chiaur, D.S., Murthy, S., Cenciarelli, C., Parks, W., Loda, M., Inghirami, G., Demetrick, D., and Pagano, M. (2000). Five human genes encoding F-box proteins: chromosome mapping and analysis in human tumors. Cytogenet Cell Genet *88*, 255-258.

Cho, H.H., Song, J.S., Yu, J.M., Yu, S.S., Choi, S.J., Kim, D.H., and Jung, J.S. (2008). Differential effect of NF-kappaB activity on beta-catenin/Tcf pathway in various cancer cells. FEBS Lett *582*, 616-622.

Cho, K.-H., Shin, S.-Y., Lee, H.-W., and Wolkenhauer, O. (2003). Investigations Into the Analysis and Modeling of the TNFα-Mediated NF-κB-Signaling Pathway. Genome Research *13*, 2413-2422.

Cho, K.H., Baek, S., and Sung, M.H. (2006). Wnt pathway mutations selected by optimal beta-catenin signaling for tumorigenesis. FEBS Lett *580*, 3665-3670.

Clague, M.J., and Urbe, S. (2010). Ubiquitin: same molecule, different degradation pathways. Cell *143*, 682-685.

Clevers, H. (2006). Wnt/beta-catenin signaling in development and disease. Cell *127*, 469-480.

Clevers, H., and Nusse, R. (2012). Wnt/beta-Catenin Signaling and Disease. Cell *149*, 1192-1205.

Davis, M., Hatzubai, A., Andersen, J.S., Ben-Shushan, E., Fisher, G.Z., Yaron, A., Bauskin, A., Mercurio, F., Mann, M., and Ben-Neriah, Y. (2002). Pseudosubstrate regulation of the SCF(beta-TrCP) ubiquitin ligase by hnRNP-U. Genes Dev *16*, 439-451.

de Bie, P., and Ciechanover, A. (2011). Ubiquitination of E3 ligases: self-regulation of the ubiquitin system via proteolytic and non-proteolytic mechanisms. Cell Death Differ *18*, 1393-1402.

de Groof, A., te Lindert, M., van Dommelen, M., Wu, M., Willemse, M., Smift, A., Winer, M., Oerlemans, F., Pluk, H., Fransen, J., et al. (2009). Increased OXPHOS activity precedes rise in glycolytic rate in H-RasV12/E1A transformed fibroblasts that develop a Warburg phenotype. Molecular Cancer *8*, 54.

Deng, J., Miller, S.A., Wang, H.-Y., Xia, W., Wen, Y., Zhou, B.P., Li, Y., Lin, S.-Y., and Hung, M.-C. (2002). beta-catenin interacts with and inhibits NF-kappaB in human colon and breast cancer. Cancer Cell *2*, 323-334.

Deng, J., Xia, W., Miller, S.A., Wen, Y., Wang, H.Y., and Hung, M.C. (2004). Crossregulation of NF-kappaB by the APC/GSK-3beta/beta-catenin pathway. Mol Carcinog *39*, 139-146.

Dequeant, M.L., Glynn, E., Gaudenz, K., Wahl, M., Chen, J., Mushegian, A., and Pourquie, O. (2006). A complex oscillating network of signaling genes underlies the mouse segmentation clock. Science *314*, 1595-1598.

Deshaies, R.J. (1999). SCF and Cullin/Ring H2-based ubiquitin ligases. Annu Rev Cell Dev Biol *15*, 435-467.

Deshaies, R.J., and Joazeiro, C.A. (2009). RING domain E3 ubiquitin ligases. Annu Rev Biochem *78*, 399-434.

Dhooge, A., Govaerts, W., and Kuznetsov, Y.A. (2003). MATCONT: A MATLAB package for numerical bifurcation analysis of ODEs. ACM Trans Math Softw *29*, 141-164.

DiDonato, J.A., Mercurio, F., and Karin, M. (2012). NF-kappaB and the link between inflammation and cancer. Immunol Rev *246*, 379-400.

Dimitriadis, E., Trangas, T., Milatos, S., Foukas, P.G., Gioulbasanis, I., Courtis, N., Nielsen, F.C., Pandis, N., Dafni, U., Bardi, G., *et al.* (2007). Expression of oncofetal RNA-binding protein CRD-BP/IMP1 predicts clinical outcome in colon cancer. Int J Cancer *121*, 486-494.

Doble, B.W., and Woodgett, J.R. (2003). GSK-3: tricks of the trade for a multi-tasking kinase. J Cell Sci *116*, 1175-1186.

Dominguez, I., Sonenshein, G.E., and Seldin, D.C. (2009). Protein Kinase CK2 in Health and Disease. Cellular and Molecular Life Sciences *66*, 1850-1857.

Duan, Y., Liao, A.P., Kuppireddi, S., Ye, Z., Ciancio, M.J., and Sun, J. (2007). beta-Catenin activity negatively regulates bacteria-induced inflammation. Lab Invest *87*, 613-624.

Elcheva, I., Goswami, S., Noubissi, F.K., and Spiegelman, V.S. (2009). CRD-BP protects the coding region of betaTrCP1 mRNA from miR-183-mediated degradation. Mol Cell *35*, 240-246.

Feldman, R.M.R., Correll, C.C., Kaplan, K.B., and Deshaies, R.J. (1997). A Complex of Cdc4p, Skp1p, and Cdc53p/Cullin Catalyzes Ubiquitination of the Phosphorylated CDK Inhibitor Sic1p. Cell *91*, 221-230.

Finley, D. (2009). Recognition and processing of ubiquitin-protein conjugates by the proteasome. Annu Rev Biochem *78*, 477-513.

Fong, A., and Sun, S.C. (2002). Genetic evidence for the essential role of beta-transducin repeat-containing protein in the inducible processing of NF-kappa B2/p100. J Biol Chem *277*, 22111-22114.

Frankland-Searby, S., and Bhaumik, S.R. (2012). The 26S proteasome complex: An attractive target for cancer therapy. Biochimica et Biophysica Acta (BBA) - Reviews on Cancer *1825*, 64-76.

Frescas, D., and Pagano, M. (2008). Deregulated proteolysis by the F-box proteins SKP2 and beta-TrCP: tipping the scales of cancer. Nat Rev Cancer *8*, 438-449.

Fuchs, S.Y., Chen, A., Xiong, Y., Pan, Z.Q., and Ronai, Z. (1999). HOS, a human homolog of Slimb, forms an SCF complex with Skp1 and Cullin1 and targets the phosphorylation-dependent degradation of IkappaB and beta-catenin. Oncogene *18*, 2039-2046.

Fuchs, S.Y., Spiegelman, V.S., and Kumar, K.G. (2004). The many faces of beta-TrCP E3 ubiquitin ligases: reflections in the magic mirror of cancer. Oncogene *23*, 2028-2036.

Furukawa, K., and Hohmann, S. (2013). Synthetic biology: lessons from engineering yeast MAPK signalling pathways. Mol Microbiol *88*, 5-19.

Gerstein, A.V., Almeida, T.A., Zhao, G., Chess, E., Shih Ie, M., Buhler, K., Pienta, K., Rubin, M.A., Vessella, R., and Papadopoulos, N. (2002). APC/CTNNB1 (beta-catenin) pathway alterations in human prostate cancers. Genes Chromosomes Cancer *34*, 9-16.

Giles, R.H., van Es, J.H., and Clevers, H. (2003). Caught up in a Wnt storm: Wnt signaling in cancer. Biochim Biophys Acta *1653*, 1-24.

Goldberg, A.L. (2007). Functions of the proteasome: from protein degradation and immune surveillance to cancer therapy. Biochem Soc Trans *35*, 12-17.

Goldbeter, A., and Pourquie, O. (2008). Modeling the segmentation clock as a network of coupled oscillations in the Notch, Wnt and FGF signaling pathways. J Theor Biol *252*, 574-585.

Goldstein, G., Scheid, M., Hammerling, U., Schlesinger, D.H., Niall, H.D., and Boyse, E.A. (1975). Isolation of a polypeptide that has lymphocyte-differentiating properties and is probably represented universally in living cells. Proceedings of the National Academy of Sciences *72*, 11-15.

Gotschel, F., Kern, C., Lang, S., Sparna, T., Markmann, C., Schwager, J., McNelly, S., von Weizsacker, F., Laufer, S., Hecht, A., *et al.* (2008). Inhibition of GSK3 differentially modulates NF-kappaB, CREB, AP-1 and beta-catenin signaling in hepatocytes, but fails to promote TNF-alpha-induced apoptosis. Experimental cell research *314*, 1351-1366.

Groll, M., Bajorek, M., Kohler, A., Moroder, L., Rubin, D.M., Huber, R., Glickman, M.H., and Finley, D. (2000). A gated channel into the proteasome core particle. Nat Struct Biol *7*, 1062-1067.

Gu, W., Wells, A.L., Pan, F., and Singer, R.H. (2008). Feedback regulation between zipcode binding protein 1 and beta-catenin mRNAs in breast cancer cells. Mol Cell Biol *28*, 4963-4974.

Guardavaccaro, D., Kudo, Y., Boulaire, J., Barchi, M., Busino, L., Donzelli, M., Margottin-Goguet, F., Jackson, P.K., Yamasaki, L., and Pagano, M. (2003). Control of meiotic and mitotic progression by the F box protein beta-Trcp1 in vivo. Dev Cell *4*, 799-812.

Hanahan, D., and Weinberg, R.A. (2000). The hallmarks of cancer. Cell *100*, 57-70.

Hanahan, D., and Weinberg, Robert A. (2011). Hallmarks of Cancer: The Next Generation. Cell *144*, 646-674.

Haney, S., Bardwell, L., and Nie, Q. (2010). Ultrasensitive Responses and Specificity in Cell Signaling. BMC Syst Biol *4*, 119.

Harhaj, E.W., and Dixit, V.M. (2012). Regulation of NF-κB by deubiquitinases. Immunological Reviews *246*, 107-124.

Harris, T.J., and Peifer, M. (2005). Decisions, decisions: beta-catenin chooses between adhesion and transcription. Trends Cell Biol *15*, 234-237.

Hart, M., Concordet, J.P., Lassot, I., Albert, I., del los Santos, R., Durand, H., Perret, C., Rubinfeld, B., Margottin, F., Benarous, R., *et al.* (1999). The F-box protein beta-TrCP associates with phosphorylated beta-catenin and regulates its activity in the cell. Curr Biol *9*, 207-210.

Hart, M.J., de los Santos, R., Albert, I.N., Rubinfeld, B., and Polakis, P. (1998). Downregulation of beta-catenin by human Axin and its association with the APC tumor suppressor, beta-catenin and GSK3 beta. Curr Biol *8*, 573-581.

Hayden, M.S., and Ghosh, S. (2012). NF-kB, the first quarter-century: remarkable progress and outstanding questions. Genes & Development *26*, 203-234.

He, T.C., Sparks, A.B., Rago, C., Hermeking, H., Zawel, L., da Costa, L.T., Morin, P.J., Vogelstein, B., and Kinzler, K.W. (1998). Identification of c-MYC as a target of the APC pathway. Science *281*, 1509-1512.

Hecht, A., and Kemler, R. (2000). Curbing the nuclear activities of beta-catenin. Control over Wnt target gene expression. EMBO Rep *1*, 24-28.

Heinrich, R., Neel, B.G., and Rapoport, T.A. (2002). Mathematical models of protein kinase signal transduction. Mol Cell *9*, 957-970.

Heinrich, R., and Rapoport, T.A. (1974). A linear steady-state treatment of enzymatic chains. General properties, control and effector strength. Eur J Biochem *42*, 89-95.

Heinrich, R., and Schuster, S. (1996). The regulation of cellular systems (Chapman & Hall, New York).

Heissmeyer, V., Krappmann, D., Hatada, E.N., and Scheidereit, C. (2001). Shared pathways of IkappaB kinase-induced SCF(betaTrCP)-mediated ubiquitination and degradation for the NF-kappaB precursor p105 and IkappaBalpha. Mol Cell Biol *21*, 1024-1035.

Henkel, T., Machleidt, T., Alkalay, I., Kronke, M., Ben-Neriah, Y., and Baeuerle, P.A. (1993). Rapid proteolysis of I kappa B-alpha is necessary for activation of transcription factor NF-kappa B. Nature *365*, 182-185.

Hershko, A., Ciechanover, A., Heller, H., Haas, A.L., and Rose, I.A. (1980). Proposed role of ATP in protein breakdown: conjugation of protein with multiple chains of the polypeptide of ATP-dependent proteolysis. Proceedings of the National Academy of Sciences *77*, 1783-1786.

Hershko, A., Heller, H., Elias, S., and Ciechanover, A. (1983). Components of ubiquitin-protein ligase system. Resolution, affinity purification, and role in protein breakdown. Journal of Biological Chemistry *258*, 8206-8214.

Heyninck, K., and Beyaert, R. (2005). A20 inhibits NF-κB activation by dual ubiquitin-editing functions. Trends in Biochemical Sciences *30*, 1-4.

Hinz, M., Arslan, S.C., and Scheidereit, C. (2012). It takes two to tango: IkappaBs, the multifunctional partners of NF-kappaB. Immunol Rev *246*, 59-76.

Hochrainer, K., and Lipp, J. (2007). Ubiquitylation within signaling pathways in- and outside of inflammation. Thromb Haemost *97*, 370-377.

Hoeflich, K.P., Luo, J., Rubie, E.A., Tsao, M.S., Jin, O., and Woodgett, J.R. (2000). Requirement for glycogen synthase kinase-3beta in cell survival and NF-kappaB activation. Nature *406*, 86-90.

Hoffmann, A., Levchenko, A., Scott, M.L., and Baltimore, D. (2002). The IkappaB-NF-kappaB signaling module: temporal control and selective gene activation. Science *298*, 1241-1245.

Hu, B., Levine, H., and Rappel, W.J. (2011). Design principles and specificity in biological networks with cross activation. Phys Biol *8*, 026001.

Hu, D., and Yuan, J.-M. (2006). Time-Dependent Sensitivity Analysis of Biological Networks: Coupled MAPK and PI3K Signal Transduction Pathways†. The Journal of Physical Chemistry A *110*, 5361-5370.

Huibregtse, J.M., Scheffner, M., Beaudenon, S., and Howley, P.M. (1995). A family of proteins structurally and functionally related to the E6-AP ubiquitin-protein ligase. Proc Natl Acad Sci U S A *92*, 2563-2567.

Huntzicker, E.G., and Oro, A.E. (2008). Controlling hair follicle signaling pathways through polyubiquitination. The Journal of investigative dermatology *128*, 1081-1087.

Hyun Hwa, C., Hye Joon, J., Ji Sun, S., Yong Chan, B., and Jin Sup, J. (2008). Crossregulation of beta-catenin/Tcf pathway by NF-kappaB is mediated by lzts2 in human adipose tissue-derived mesenchymal stem cells. Biochim Biophys Acta *1783*, 419-428.

Ihekwaba, A.E., Broomhead, D.S., Grimley, R.L., Benson, N., and Kell, D.B. (2004). Sensitivity analysis of parameters controlling oscillatory signalling in the NF-kappaB pathway: the roles of IKK and IkappaBalpha. Syst Biol (Stevenage) *1*, 93-103.

Inestrosa, N.C., and Arenas, E. (2010). Emerging roles of Wnts in the adult nervous system. Nat Rev Neurosci *11*, 77-86.

Ioannidis, P., Trangas, T., Dimitriadis, E., Samiotaki, M., Kyriazoglou, I., Tsiapalis, C.M., Kittas, C., Agnantis, N., Nielsen, F.C., Nielsen, J., *et al.* (2001). C-MYC and IGF-II mRNA-binding protein (CRD-BP/IMP-1) in benign and malignant mesenchymal tumors. Int J Cancer *94*, 480-484.

Iwai, K. (2012). Diverse ubiquitin signaling in NF-kappaB activation. Trends Cell Biol *22*, 355-364.

Jiang, J., and Struhl, G. (1998). Regulation of the Hedgehog and Wingless signalling pathways by the F-box/WD40-repeat protein Slimb. Nature *391*, 493-496.

Joo, J., Plimpton, S., Martin, S., Swiler, L., and Faulon, J.L. (2007). Sensitivity analysis of a computational model of the IKK NF-kappaB IkappaBalpha A20 signal transduction network. Ann N Y Acad Sci *1115*, 221-239.

Joyce, D., Albanese, C., Steer, J., Fu, M., Bouzahzah, B., and Pestell, R.G. (2001). NF-kappaB and cell-cycle regulation: the cyclin connection. Cytokine Growth Factor Rev *12*, 73-90.

Kacser, H., and Burns, J.A. (1973). The control of flux. Symposia of the Society for Experimental Biology *27*, 65-104.

Kanarek, N., and Ben-Neriah, Y. (2012). Regulation of NF-kappaB by ubiquitination and degradation of the IkappaBs. Immunol Rev *246*, 77-94.

Kanarek, N., Horwitz, E., Mayan, I., Leshets, M., Cojocaru, G., Davis, M., Tsuberi, B.Z., Pikarsky, E., Pagano, M., and Ben-Neriah, Y. (2010). Spermatogenesis rescue in a mouse deficient for the ubiquitin ligase SCF{beta}-TrCP by single substrate depletion. Genes Dev *24*, 470-477.

Karin, M. (2009). NF-κB as a Critical Link Between Inflammation and Cancer. Cold Spring Harb Perspect Biol *1*.

Kendall, M.G. (1938). A new measure of rank corellation. Biometrika *30*, 81-93.

Kholodenko, B.N., Hoek, J.B., Westerhoff, H.V., and Brown, G.C. (1997). Quantification of information transfer via cellular signal transduction pathways. FEBS letters *414*, 430-434.

Kim, C.J., Song, J.H., Cho, Y.G., Kim, Y.S., Kim, S.Y., Nam, S.W., Yoo, N.J., Lee, J.Y., and Park, W.S. (2007a). Somatic mutations of the beta-TrCP gene in gastric cancer. APMIS *115*, 127-133.

Kim, D., Kolch, W., and Cho, K.H. (2009). Multiple roles of the NF-kappaB signaling pathway regulated by coupled negative feedback circuits. FASEB J *23*, 2796-2802.

Kim, D., Rath, O., Kolch, W., and Cho, K.H. (2007b). A hidden oncogenic positive feedback loop caused by crosstalk between Wnt and ERK pathways. Oncogene *26*, 4571-4579.

Kimelman, D., and Xu, W. (2006). beta-catenin destruction complex: insights and questions from a structural perspective. Oncogene *25*, 7482-7491.

Kirkpatrick, S., Gelatt, C.D., Jr., and Vecchi, M.P. (1983). Optimization by simulated annealing. Science *220*, 671-680.

Kitagawa, M., Hatakeyama, S., Shirane, M., Matsumoto, M., Ishida, N., Hattori, K., Nakamichi, I., Kikuchi, A., and Nakayama, K. (1999). An F-box protein, FWD1, mediates ubiquitin-dependent proteolysis of beta-catenin. EMBO J *18*, 2401-2410.

Klaus, A., and Birchmeier, W. (2008). Wnt signalling and its impact on development and cancer. Nat Rev Cancer *8*, 387-398.

Klinger, B., Sieber, A., Fritsche-Guenther, R., Witzel, F., Berry, L., Schumacher, D., Yan, Y., Durek, P., Merchant, M., Schafer, R.*, et al.* (2013). Network quantification of EGFR signaling unveils potential for targeted combination therapy. Mol Syst Biol *9*, 673.

Klipp, E. (2009). Timing matters. FEBS Letters *583*, 4013-4018.

Klipp, E., and Liebermeister, W. (2006). Mathematical modeling of intracellular signaling pathways. BMC Neurosci *7 Suppl 1*, S10.

Kofahl, B., and Klipp, E. (2004). Modelling the dynamics of the yeast pheromone pathway. Yeast *21*, 831-850.

Kofahl, B., and Wolf, J. (2010). Mathematical modelling of Wnt/beta-catenin signalling. Biochem Soc Trans *38*, 1281-1285.

Koike, J., Sagara, N., Kirikoshi, H., Takagi, A., Miwa, T., Hirai, M., and Katoh, M. (2000). Molecular cloning and genomic structure of the betaTRCP2 gene on chromosome 5q35.1. Biochem Biophys Res Commun *269*, 103-109.

Komarova, N.L., Zou, X., Nie, Q., and Bardwell, L. (2005). A theoretical framework for specificity in cell signaling. Mol Syst Biol *1*, 2005 0023.

Krishna, S., Jensen, M.H., and Sneppen, K. (2006). Minimal model of spiky oscillations in NF-kappaB signaling. Proc Natl Acad Sci U S A *103*, 10840-10845.

Kroll, M., Margottin, F., Kohl, A., Renard, P., Durand, H., Concordet, J.P., Bachelerie, F., Arenzana-Seisdedos, F., and Benarous, R. (1999). Inducible degradation of IkappaBalpha by the proteasome requires interaction with the F-box protein h-betaTrCP. J Biol Chem *274*, 7941-7945.

Kruger, R., and Heinrich, R. (2004). Model reduction and analysis of robustness for the Wnt/beta-catenin signal transduction pathway. Genome Inform *15*, 138-148.

Kuhl, M., Sheldahl, L.C., Park, M., Miller, J.R., and Moon, R.T. (2000). The Wnt/Ca2+ pathway: a new vertebrate Wnt signaling pathway takes shape. Trends Genet *16*, 279-283.

Lamberti, C., Lin, K.M., Yamamoto, Y., Verma, U., Verma, I.M., Byers, S., and Gaynor, R.B. (2001). Regulation of beta-catenin function by the IkappaB kinases. J Biol Chem *276*, 42276-42286.

Lang, F., Ritter, M., Woll, E., Weiss, H., Haussinger, D., Hoflacher, J., Maly, K., and Grunicke, H. (1992). Altered cell volume regulation in ras oncogene expressing NIH fibroblasts. Pflugers Arch *420*, 424-427.

Lang, V., Janzen, J., Fischer, G.Z., Soneji, Y., Beinke, S., Salmeron, A., Allen, H., Hay, R.T., Ben-Neriah, Y., and Ley, S.C. (2003). betaTrCP-mediated proteolysis of NF-kappaB1 p105 requires phosphorylation of p105 serines 927 and 932. Mol Cell Biol *23*, 402-413.

Latres, E., Chiaur, D.S., and Pagano, M. (1999). The human F box protein beta-Trcp associates with the Cul1/Skp1 complex and regulates the stability of beta-catenin. Oncogene *18*, 849-854.

Lau, A.W., Fukushima, H., and Wei, W. (2012). The Fbw7 and betaTRCP E3 ubiquitin ligases and their roles in tumorigenesis. Frontiers in bioscience : a journal and virtual library *17*, 2197-2212.

Lee, E., Salic, A., Kruger, R., Heinrich, R., and Kirschner, M.W. (2003). The roles of APC and Axin derived from experimental and theoretical analysis of the Wnt pathway. PLoS Biol *1*, E10.

Li, Y., Gazdoiu, S., Pan, Z.Q., and Fuchs, S.Y. (2004). Stability of homologue of Slimb F-box protein is regulated by availability of its substrate. J Biol Chem *279*, 11074-11080.

Lipniacki, T., Paszek, P., Brasier, A.R., Luxon, B., and Kimmel, M. (2004). Mathematical model of NF-kappaB regulatory module. J Theor Biol *228*, 195-215.

Liu, C., Kato, Y., Zhang, Z., Do, V.M., Yankner, B.A., and He, X. (1999). beta-Trcp couples beta-catenin phosphorylation-degradation and regulates Xenopus axis formation. Proc Natl Acad Sci U S A *96*, 6273-6278.

Llorens, M., Nuno, J.C., Rodriguez, Y., Melendez-Hevia, E., and Montero, F. (1999). Generalization of the theory of transition times in metabolic pathways: a geometrical approach. Biophys J *77*, 23-36.

Luckert, K., Gujral, T.S., Chan, M., Sevecka, M., Joos, T.O., Sorger, P.K., MacBeath, G., and Potz, O. (2012). A Dual Array-Based Approach to Assess the Abundance and Posttranslational Modification State of Signaling Proteins. Sci Signal *5*, pl1-.

MacDonald, B.T., and He, X. (2012). Frizzled and LRP5/6 Receptors for Wnt/beta-Catenin Signaling. Cold Spring Harb Perspect Biol *4*.

MacDonald, B.T., Tamai, K., and He, X. (2009). Wnt/beta-catenin signaling: components, mechanisms, and diseases. Dev Cell *17*, 9-26.

Major, M.B., Camp, N.D., Berndt, J.D., Yi, X., Goldenberg, S.J., Hubbert, C., Biechele, T.L., Gingras, A.C., Zheng, N., Maccoss, M.J._, et al._ (2007). Wilms tumor suppressor WTX negatively regulates WNT/beta-catenin signaling. Science *316*, 1043-1046.

Maniatis, T. (1999). A ubiquitin ligase complex essential for the NF-kappaB, Wnt/Wingless, and Hedgehog signaling pathways. Genes Dev *13*, 505-510.

Margottin, F., Bour, S.P., Durand, H., Selig, L., Benichou, S., Richard, V., Thomas, D., Strebel, K., and Benarous, R. (1998). A novel human WD protein, h-beta TrCp, that interacts with HIV-1 Vpu connects CD4 to the ER degradation pathway through an F-box motif. Mol Cell *1*, 565-574.

Mathes, E., O'Dea, E.L., Hoffmann, A., and Ghosh, G. (2008). NF-[kappa]B dictates the degradation pathway of I[kappa]B[alpha]. EMBO J *27*, 1357-1367.

McNeill, H., and Woodgett, J.R. (2010). When pathways collide: collaboration and connivance among signalling proteins in development. Nat Rev Mol Cell Biol *11*, 404-413.

Merrill, B.J. (2012). Wnt Pathway Regulation of Embryonic Stem Cell Self-Renewal. Cold Spring Harb Perspect Biol *4*.

Mirams, G.R., Byrne, H.M., and King, J.R. (2010). A multiple timescale analysis of a mathematical model of the Wnt/beta-catenin signalling pathway. J Math Biol *60*, 131-160.

Muerkoster, S., Arlt, A., Sipos, B., Witt, M., Grossmann, M., Kloppel, G., Kalthoff, H., Folsch, U.R., and Schafer, H. (2005). Increased expression of the E3-ubiquitin ligase receptor subunit betaTRCP1 relates to constitutive nuclear factor-kappaB activation and chemoresistance in pancreatic carcinoma cells. Cancer Res *65*, 1316-1324.

Nakajima, H., Fujiwara, H., Furuichi, Y., Tanaka, K., and Shimbara, N. (2008). A novel small-molecule inhibitor of NF-κB signaling. Biochemical and Biophysical Research Communications *368*, 1007-1013.

Nakayama, K., Hatakeyama, S., Maruyama, S., Kikuchi, A., Onoe, K., Good, R.A., and Nakayama, K.I. (2003). Impaired degradation of inhibitory subunit of NF-kappa B (I kappa B) and beta-catenin as a result of targeted disruption of the beta-TrCP1 gene. Proc Natl Acad Sci U S A *100*, 8752-8757.

Navon, A., and Goldberg, A.L. (2001). Proteins Are Unfolded on the Surface of the ATPase Ring before Transport into the Proteasome. Molecular Cell *8*, 1339-1349.

Nejak-Bowen, K., Kikuchi, A., and Monga, S.P.S. (2012). Beta-catenin-NF-κB interactions in murine hepatocytes: A complex to die for. Hepatology, n/a-n/a.

Nelson, D.E., Horton, C.A., See, V., Johnson, J.R., Nelson, G., Spiller, D.G., Kell, D.B., and White, M.R.H. (2005). Response to Comment on "Oscillations in NF-κB Signaling Control the Dynamics of Gene Expression". Science *308*, 52.

Nelson, D.E., Ihekwaba, A.E.C., Elliott, M., Johnson, J.R., Gibney, C.A., Foreman, B.E., Nelson, G., See, V., Horton, C.A., Spiller, D.G., *et al.* (2004). Oscillations in NF-kB Signaling Control the Dynamics of Gene Expression. Science *306*, 704-708.

Nguyen, L.K., Cavadas, M.A., Scholz, C.C., Fitzpatrick, S.F., Bruning, U., Cummins, E.P., Tambuwala, M.M., Manresa, M.C., Kholodenko, B.N., Taylor, C.T., *et al.* (2013). A dynamic model of the hypoxia-inducible factor 1alpha (HIF-1alpha) network. Journal of Cell Science *126*, 1454-1463.

Niehrs, C. (2012). The complex world of WNT receptor signalling. Nat Rev Mol Cell Biol *13*, 767-779.

Nikolov, S., Vera, J., Rath, O., Kolch, W., and Wolkenhauer, O. (2009). Role of inhibitory proteins as modulators of oscillations in NFB signalling. IET systems biology *3*, 59-76.

Noubissi, F.K., Elcheva, I., Bhatia, N., Shakoori, A., Ougolkov, A., Liu, J., Minamoto, T., Ross, J., Fuchs, S.Y., and Spiegelman, V.S. (2006). CRD-BP mediates stabilization of betaTrCP1 and c-myc mRNA in response to beta-catenin signalling. Nature *441*, 898-901.

Nusse, R., and Varmus, H. (2012). Three decades of Wnts: a personal perspective on how a scientific field developed. EMBO J *31*, 2670-2684.

O'Dea, E., and Hoffmann, A. (2010). The regulatory logic of the NF-kappaB signaling system. Cold Spring Harb Perspect Biol *2*, a000216.

Oeckinghaus, A., and Ghosh, S. (2009). The NF-kB Family of Transcription Factors and Its Regulation. Cold Spring Harb Perspect Biol *1*.

Orian, A., Gonen, H., Bercovich, B., Fajerman, I., Eytan, E., Israel, A., Mercurio, F., Iwai, K., Schwartz, A.L., and Ciechanover, A. (2000). SCF(beta)(-TrCP) ubiquitin ligase-mediated processing of NF-kappaB p105 requires phosphorylation of its C-terminus by IkappaB kinase. EMBO J *19*, 2580-2591.

Ougolkov, A., Zhang, B., Yamashita, K., Bilim, V., Mai, M., Fuchs, S.Y., and Minamoto, T. (2004). Associations among beta-TrCP, an E3 ubiquitin ligase receptor, beta-catenin, and NF-kappaB in colorectal cancer. J Natl Cancer Inst *96*, 1161-1170.

Pahl, H.L. (1999). Activators and target genes of Rel/NF-kappaB transcription factors. Oncogene *18*, 6853-6866.

Petroski, M.D., and Deshaies, R.J. (2005). Function and regulation of cullin-RING ubiquitin ligases. Nat Rev Mol Cell Biol *6*, 9-20.

Pickart, C.M. (2004). Back to the future with ubiquitin. Cell *116*, 181-190.

Ravid, T., and Hochstrasser, M. (2008). Diversity of degradation signals in the ubiquitin-proteasome system. Nat Rev Mol Cell Biol *9*, 679-690.

Reder, C. (1988). Metabolic control theory: a structural approach. Journal of theoretical biology *135*, 175-201.

Reifenberger, J., Knobbe, C.B., Wolter, M., Blaschke, B., Schulte, K.W., Pietsch, T., Ruzicka, T., and Reifenberger, G. (2002). Molecular genetic analysis of malignant melanomas for aberrations of the WNT signaling pathway genes CTNNB1, APC, ICAT and BTRC. Int J Cancer *100*, 549-556.

Saeki, Y., and Tanaka, K. (2012). Assembly and function of the proteasome. Methods Mol Biol *832*, 315-337.

Saito, H. (2010). Regulation of cross-talk in yeast MAPK signaling pathways. Current Opinion in Microbiology *13*, 677-683.

Saitoh, T., and Katoh, M. (2001). Expression profiles of betaTRCP1 and betaTRCP2, and mutation analysis of betaTRCP2 in gastric cancer. Int J Oncol *18*, 959-964.

Sanchez, J.F., Sniderhan, L.F., Williamson, A.L., Fan, S., Chakraborty-Sett, S., and Maggirwar, S.B. (2003). Glycogen synthase kinase 3beta-mediated apoptosis of primary cortical astrocytes involves inhibition of nuclear factor kappaB signaling. Mol Cell Biol *23*, 4649-4662.

Schaber, J., Baltanas, R., Bush, A., Klipp, E., and Colman-Lerner, A. (2012). Modelling reveals novel roles of two parallel signalling pathways and homeostatic feedbacks in yeast. Mol Syst Biol *8*, 622.

Schaber, J., Kofahl, B., Kowald, A., and Klipp, E. (2006). A modelling approach to quantify dynamic crosstalk between the pheromone and the starvation pathway in baker's yeast. FEBS Journal *273*, 3520-3533.

Scheidereit, C. (2006). IkappaB kinase complexes: gateways to NF-kappaB activation and transcription. Oncogene *25*, 6685-6705.

Schmitz, Y., Rateitschak, K., and Wolkenhauer, O. (2013). Analysing the impact of nucleo-cytoplasmic shuttling of β-catenin and its antagonists APC, Axin and GSK3 on Wnt/β-catenin signalling. Cellular Signalling.

Schmitz, Y., Wolkenhauer, O., and Rateitschak, K. (2011). Nucleo-cytoplasmic shuttling of APC can maximize β-catenin/TCF concentration. Journal of Theoretical Biology *279*, 132-142.

Schmukle, A.C., and Walczak, H. (2012). No one can whistle a symphony alone – how different ubiquitin linkages cooperate to orchestrate NF-κB activity. Journal of Cell Science *125*, 549-559.

Schwabe, R.F., and Brenner, D.A. (2002). Role of glycogen synthase kinase-3 in TNF-alpha-induced NF-kappaB activation and apoptosis in hepatocytes. Am J Physiol Gastrointest Liver Physiol *283*, G204-211.

Schwanhausser, B., Busse, D., Li, N., Dittmar, G., Schuchhardt, J., Wolf, J., Chen, W., and Selbach, M. (2011). Global quantification of mammalian gene expression control. Nature *473*, 337-342.

Schwartz, M.A., and Madhani, H.D. (2004). Principles of MAP kinase signaling specificity in Saccharomyces cerevisiae. Annu Rev Genet *38*, 725-748.

Seaton, D.D., and Krishnan, J. (2011). The coupling of pathways and processes through shared components. BMC Syst Biol *5*, 103.

Sen, R., and Smale, S.T. (2010). Selectivity of the NF-{kappa}B response. Cold Spring Harb Perspect Biol *2*, a000257.

Seo, E., Kim, H., Kim, R., Yun, S., Kim, M., Han, J.K., Costantini, F., and Jho, E.H. (2009). Multiple isoforms of beta-TrCP display differential activities in the regulation of Wnt signaling. Cell Signal *21*, 43-51.

Sharp, P.M., and Li, W.-H. (1987). Molecular evolution of ubiquitin genes. Trends in Ecology & Evolution *2*, 328-332.

Shih, V.F.-S., Kearns, J.D., Basak, S., Savinova, O.V., Ghosh, G., and Hoffmann, A. (2009). Kinetic control of negative feedback regulators of NF-κB/RelA determines their pathogen- and cytokine-receptor signaling specificity. Proceedings of the National Academy of Sciences *106*, 9619-9624.

Shirane, M., Hatakeyama, S., Hattori, K., Nakayama, K., and Nakayama, K.-i. (1999). Common Pathway for the Ubiquitination of IκBα, IκBβ, and IκBε Mediated by the F-Box Protein FWD1. Journal of Biological Chemistry *274*, 28169-28174.

Shtutman, M., Zhurinsky, J., Simcha, I., Albanese, C., D'Amico, M., Pestell, R., and Ben-Ze'ev, A. (1999). The cyclin D1 gene is a target of the beta-catenin/LEF-1 pathway. Proc Natl Acad Sci U S A *96*, 5522-5527.

Sims, C.E., and Allbritton, N.L. (2007). Analysis of single mammalian cells on-chip. Lab on a Chip *7*, 423-440.

Skowyra, D., Craig, K.L., Tyers, M., Elledge, S.J., and Harper, J.W. (1997). F-Box Proteins Are Receptors that Recruit Phosphorylated Substrates to the SCF Ubiquitin-Ligase Complex. Cell *91*, 209-219.

Skowyra, D., Koepp, D.M., Kamura, T., Conrad, M.N., Conaway, R.C., Conaway, J.W., Elledge, S.J., and Harper, J.W. (1999). Reconstitution of G1 Cyclin Ubiquitination with Complexes Containing SCFGrr1 and Rbx1. Science *284*, 662-665.

Smith, T.F., Gaitatzes, C., Saxena, K., and Neer, E.J. (1999). The WD repeat: a common architecture for diverse functions. Trends in Biochemical Sciences *24*, 181-185.

Soldatenkov, V.A., Dritschilo, A., Ronai, Z., and Fuchs, S.Y. (1999). Inhibition of homologue of Slimb (HOS) function sensitizes human melanoma cells for apoptosis. Cancer Res *59*, 5085-5088.

Spencer, E., Jiang, J., and Chen, Z.J. (1999). Signal-induced ubiquitination of IkappaBalpha by the F-box protein Slimb/beta-TrCP. Genes Dev *13*, 284-294.

Spevak, W., Keiper, B.D., Stratowa, C., and CastaÃ±Ã³n, M.J. (1993). Saccharomyces cerevisiae cdc15 mutants arrested at a late stage in anaphase are rescued by Xenopus cDNAs encoding N-ras or a protein with beta-transducin repeats. Molecular and Cellular Biology *13*, 4953-4966.

Spiegelman, V.S., Slaga, T.J., Pagano, M., Minamoto, T., Ronai, Z., and Fuchs, S.Y. (2000). Wnt/beta-catenin signaling induces the expression and activity of betaTrCP ubiquitin ligase receptor. Mol Cell *5*, 877-882.

Spiegelman, V.S., Stavropoulos, P., Latres, E., Pagano, M., Ronai, Z., Slaga, T.J., and Fuchs, S.Y. (2001). Induction of beta-transducin repeat-containing protein by JNK signaling and its role in the activation of NF-kappaB. J Biol Chem *276*, 27152-27158.

Spiegelman, V.S., Tang, W., Chan, A.M., Igarashi, M., Aaronson, S.A., Sassoon, D.A., Katoh, M., Slaga, T.J., and Fuchs, S.Y. (2002a). Induction of homologue of Slimb ubiquitin ligase receptor by mitogen signaling. J Biol Chem *277*, 36624-36630.

Spiegelman, V.S., Tang, W., Katoh, M., Slaga, T.J., and Fuchs, S.Y. (2002b). Inhibition of HOS expression and activities by Wnt pathway. Oncogene *21*, 856-860.

Stadeli, R., Hoffmans, R., and Basler, K. (2006). Transcription under the control of nuclear Arm/beta-catenin. Curr Biol *16*, R378-385.

Stamos, J.L., and Weis, W.I. (2013). The beta-Catenin Destruction Complex. Cold Spring Harb Perspect Biol *5*.

Staudt, L.M. (2010). Oncogenic Activation of NF-κB. Cold Spring Harb Perspect Biol *2*.

Su, Y., Fu, C., Ishikawa, S., Stella, A., Kojima, M., Shitoh, K., Schreiber, E.M., Day, B.W., and Liu, B. (2008). APC is essential for targeting phosphorylated beta-catenin to the SCFbeta-TrCP ubiquitin ligase. Mol Cell *32*, 652-661.

Sun, J., Hobert, M.E., Duan, Y., Rao, A.S., He, T.C., Chang, E.B., and Madara, J.L. (2005). Crosstalk between NF-kappaB and beta-catenin pathways in bacterial-colonized intestinal epithelial cells. Am J Physiol Gastrointest Liver Physiol *289*, G129-137.

Sun, S.-C. (2012). The noncanonical NF-κB pathway. Immunological Reviews *246*, 125-140.

Suzuki, H., Chiba, T., Kobayashi, M., Takeuchi, M., Suzuki, T., Ichiyama, A., Ikenoue, T., Omata, M., Furuichi, K., and Tanaka, K. (1999). IkappaBalpha ubiquitination is catalyzed by an SCF-like complex containing Skp1, cullin-1, and two F-box/WD40-repeat proteins, betaTrCP1 and betaTrCP2. Biochem Biophys Res Commun *256*, 127-132.

Suzuki, H., Chiba, T., Suzuki, T., Fujita, T., Ikenoue, T., Omata, M., Furuichi, K., Shikama, H., and Tanaka, K. (2000). Homodimer of two F-box proteins betaTrCP1 or betaTrCP2 binds to IkappaBalpha for signal-dependent ubiquitination. J Biol Chem *275*, 2877-2884.

Tan, C.W., Gardiner, B.S., Hirokawa, Y., Layton, M.J., Smith, D.W., and Burgess, A.W. (2012). Wnt signalling pathway parameters for mammalian cells. PLoS One 7, e31882.

Tan, P., Fuchs, S.Y., Chen, A., Wu, K., Gomez, C., Ronai, Z.e., and Pan, Z.-Q. (1999). Recruitment of a ROC1â€"CUL1 Ubiquitin Ligase by Skp1 and HOS to Catalyze the Ubiquitination of IÎ°BÎ±. Molecular Cell 3, 527-533.

Tauriello, D.V., and Maurice, M.M. (2010). The various roles of ubiquitin in Wnt pathway regulation. Cell Cycle 9, 3700-3709.

Tay, S., Hughey, J.J., Lee, T.K., Lipniacki, T., Quake, S.R., and Covert, M.W. (2010). Single-cell NF-kappaB dynamics reveal digital activation and analogue information processing. Nature.

Thalhauser, C.J., and Komarova, N.L. (2009). Specificity and robustness of the mammalian MAPK-IEG network. Biophys J 96, 3471-3482.

Thyssen, G., Li, T.H., Lehmann, L., Zhuo, M., Sharma, M., and Sun, Z. (2006). LZTS2 is a novel beta-catenin-interacting protein and regulates the nuclear export of beta-catenin. Mol Cell Biol 26, 8857-8867.

Tsuchiya, Y., Asano, T., Nakayama, K., Kato Jr, T., Karin, M., and Kamata, H. (2010). Nuclear IKKbeta Is an Adaptor Protein for IkBalpha Ubiquitination and Degradation in UV-Induced NF-kB Activation. Molecular Cell 39, 570-582.

Tyson, J.J., Chen, K.C., and Novak, B. (2003). Sniffers, buzzers, toggles and blinkers: dynamics of regulatory and signaling pathways in the cell. Curr Opin Cell Biol 15, 221-231.

Valenta, T., Hausmann, G., and Basler, K. (2012). The many faces and functions of beta-catenin. EMBO J 31, 2714-2736.

van Amerongen, R., Mikels, A., and Nusse, R. (2008). Alternative wnt signaling is initiated by distinct receptors. Sci Signal 1, re9.

van Leeuwen, I.M., Byrne, H.M., Jensen, O.E., and King, J.R. (2007). Elucidating the interactions between the adhesive and transcriptional functions of beta-catenin in normal and cancerous cells. J Theor Biol 247, 77-102.

Veeman, M.T., Axelrod, J.D., and Moon, R.T. (2003). A second canon. Functions and mechanisms of beta-catenin-independent Wnt signaling. Dev Cell 5, 367-377.

Wang, C.Y., Mayo, M.W., and Baldwin, A.S., Jr. (1996). TNF- and cancer therapy-induced apoptosis: potentiation by inhibition of NF-kappaB. Science 274, 784-787.

Wang, X., Adhikari, N., Li, Q., Guan, Z., and Hall, J.L. (2004). The role of [beta]-transducin repeat-containing protein ([beta]-TrCP) in the regulation of NF-[kappa]B in vascular smooth muscle cells. Arterioscler Thromb Vasc Biol 24, 85-90.

Wawra, C., Kuhl, M., and Kestler, H.A. (2007). Extended analyses of the Wnt/beta-catenin pathway: robustness and oscillatory behaviour. FEBS Lett 581, 4043-4048.

Werner, S.L., Barken, D., and Hoffmann, A. (2005). Stimulus Specificity of Gene Expression Programs Determined by Temporal Control of IKK Activity. Science 309, 1857-1861.

Wertz, I.E., O'Rourke, K.M., Zhou, H., Eby, M., Aravind, L., Seshagiri, S., Wu, P., Wiesmann, C., Baker, R., Boone, D.L., et al. (2004). De-ubiquitination and ubiquitin ligase domains of A20 downregulate NF-[kappa]B signalling. Nature 430, 694-699.

Willert, K., and Nusse, R. (2012). Wnt Proteins. Cold Spring Harb Perspect Biol *4*.

Winston, J.T., Strack, P., Beer-Romero, P., Chu, C.Y., Elledge, S.J., and Harper, J.W. (1999). The SCFbeta-TRCP-ubiquitin ligase complex associates specifically with phosphorylated destruction motifs in IkappaBalpha and beta-catenin and stimulates IkappaBalpha ubiquitination in vitro. Genes Dev *13*, 270-283.

Wodke, J.A., Puchalka, J., Lluch-Senar, M., Marcos, J., Yus, E., Godinho, M., Gutierrez-Gallego, R., dos Santos, V.A., Serrano, L., Klipp, E., *et al.* (2013). Dissecting the energy metabolism in Mycoplasma pneumoniae through genome-scale metabolic modeling. Mol Syst Biol *9*, 653.

Wolkenhauer, O., Sreenath, S.N., Wellstead, P., Ullah, M., and Cho, K.H. (2005). A systems- and signal-oriented approach to intracellular dynamics. Biochemical Society transactions *33*, 507-515.

Wolter, M., Scharwachter, C., Reifenberger, J., Koch, A., Pietsch, T., and Reifenberger, G. (2003). Absence of detectable alterations in the putative tumor suppressor gene BTRC in cerebellar medulloblastomas and cutaneous basal cell carcinomas. Acta Neuropathol *106*, 287-290.

Wu, C., and Ghosh, S. (1999). beta-TrCP mediates the signal-induced ubiquitination of IkappaBbeta. J Biol Chem *274*, 29591-29594.

Wu, G., Xu, G., Schulman, B.A., Jeffrey, P.D., Harper, J.W., and Pavletich, N.P. (2003). Structure of a beta-TrCP1-Skp1-beta-catenin complex: destruction motif binding and lysine specificity of the SCF(beta-TrCP1) ubiquitin ligase. Mol Cell *11*, 1445-1456.

Yang, L., Chen, H., and Qwarnstrom, E. (2001). Degradation of IkappaBalpha is limited by a postphosphorylation/ubiquitination event. Biochem Biophys Res Commun *285*, 603-608.

Yaron, A., Gonen, H., Alkalay, I., Hatzubai, A., Jung, S., Beyth, S., Mercurio, F., Manning, A.M., Ciechanover, A., and Ben-Neriah, Y. (1997). Inhibition of NF-[kappa]B cellular function via specific targeting of the I[kappa]B-ubiquitin ligase. EMBO J *16*, 6486-6494.

Yaron, A., Hatzubai, A., Davis, M., Lavon, I., Amit, S., Manning, A.M., Andersen, J.S., Mann, M., Mercurio, F., and Ben-Neriah, Y. (1998). Identification of the receptor component of the I[kappa]B[alpha]-ubiquitin ligase. Nature *396*, 590-594.

Yi, T.M., Andrews, B.W., and Iglesias, P.A. (2007). Control Analysis of Bacterial Chemotaxis Signaling. In Methods in Enzymology, B.R.C. Melvin I. Simon, and C. Alexandrine, eds. (Academic Press), pp. 123-140.

Yun, K., Choi, Y.D., Nam, J.H., Park, Z., and Im, S.-H. (2007). NF-kB regulates Lef1 gene expression in chondrocytes. Biochemical and Biophysical Research Communications *357*, 589-595.

Zhang, H., Zhang, Y., Ng, S.S., Ren, F., Wang, Y., Duan, Y., Chen, L., Zhai, Y., Guo, Q., and Chang, Z. (2010). Dishevelled-DEP domain interacting protein (DDIP) inhibits Wnt signaling by promoting TCF4 degradation and disrupting the TCF4/beta-catenin complex. Cell Signal *22*, 1753-1760.

Zhang, M., Yan, Y., Lim, Y.B., Tang, D., Xie, R., Chen, A., Tai, P., Harris, S.E., Xing, L., Qin, Y.X., *et al.* (2009). BMP-2 modulates beta-catenin signaling through stimulation of Lrp5 expression and inhibition of beta-TrCP expression in osteoblasts. J Cell Biochem *108*, 896-905.

Zimmerman, Z.F., Moon, R.T., and Chien, A.J. (2012). Targeting Wnt pathways in disease. Cold Spring Harb Perspect Biol *4*.

List of Publications

Manuscripts in preparation:

Benary U, Hecht A, Wolf J: "Dissecting the differential impact of FWD1 and HOS feedback on Wnt/β-catenin signalling."

Benary U, Wolf J: "Mathematical modelling of β-TrCP-dependent crosstalk between canonical NF-κB signalling and Wnt/β-catenin signalling."

Publications during PhD study:

Benary U, Kofahl B, Hecht A, Wolf J: "Modelling Wnt/ß-catenin target gene expression in APC and Wnt gradients under wild type and mutant conditions." Frontiers in Physiology. 2013 Feb 25; 4(21).

Publications before PhD study:

Podtschaske M, **Benary** U, Zwinger S, Höfer T, Radbruch A, Baumgrass R: "Digital NFATc2 activation per cell transforms graded T cell receptor activation into an all-or-none IL-2 expression." PLoS ONE. 2007 Sep 26; 2(9):e935.

Diploma Thesis:

"Kinetics of Transcription Factor NFAT1 and Cytokine Expression in Stimulated Human T-Helper Cells." German Rheumatology Research Center (DRFZ) Berlin, 2006.

Berlin, den 26. Juli 2013

Uwe Benary

Acknowledgements

First of all, I would like to thank all present and former members of Dr. Jana Wolf's group at the Max-Delbrück-Center (MDC) in Berlin-Buch. Especially, my supervisor Dr. Jana Wolf for her support, guidance, and patience throughout the past five years of my PhD. Working in her group has been full of great experiences and a lot of fun. The members of her group have done me many favours for which I am much obliged. In particular, I thank Bente, Dorothea, and Janina for their suggestions concerning this manuscript.

I would also like to thank my supervisor at the Humboldt University of Berlin Prof. Dr. Dr. h.c. Edda Klipp for inviting me to frequently present and discuss my research results in her group. I am also grateful for her kind and immediate help whenever I struggled with administrative issues at university.

In addition, I would like to acknowledge the pleasant and productive collaboration with Prof. Dr. Andreas Hecht from the University of Freiburg in several side-projects during my PhD. I furthermore appreciate the help from the members of the PhD committee of the MDC (especially, Prof. Dr. Scheidereit and Dr. Kempa) for their suggestions and discussions. At last, I have to acknowledge that the studies in this thesis were funded by the MSBN project within the Helmholtz Alliance on Systems Biology funded by the Initiative and Networking Fund of the Helmholtz Association.